Developed and Published
by
AIMS Education Foundation

This book contains materials developed by the AIMS Education Foundation. **AIMS** (**A**ctivities **I**ntegrating **M**athematics and **S**cience) began in 1981 with a grant from the National Science Foundation. The non-profit AIMS Education Foundation publishes hands-on instructional materials that build conceptual understanding. The foundation also sponsors a national program of professional development through which educators may gain expertise in teaching math and science.

Copyright © 2009 by the AIMS Education Foundation

All rights reserved. No part of this book or compact disc may be reproduced or transmitted in any form or by any means—graphic, electronic, or mechanical, including photocopying, taping, or information storage/retrieval systems—except as noted below.

- A person or school purchasing this AIMS publication is hereby granted permission to make up to 200 copies of any portion of it, provided these copies will be used for educational purposes and only at that school site. The accompanying compact disc contains all the printable pages for the publication. No files on the compact disc may be shared or altered by any means.

  Adobe® Acrobat® Reader (version 5.0 or higher) is required to access files on the compact disc. Adobe® Acrobat® Reader can be downloaded at no charge at adobe.com.

- An educator providing a professional-development workshop is granted permission to make enough copies of student activity sheets to use the lessons one time with one group.

AIMS users may purchase unlimited duplication rights for making additional copies or for use at more than one school site. Contact Duplication Rights.

AIMS Education Foundation
P.O. Box 8120, Fresno, CA 93747-8120 • 888.733.2467 • aimsedu.org

ISBN 978-1-60519-007-5

Printed in the United States of America

SHAPES, SOLIDS, AND MORE
TABLE OF CONTENTS

Shapes, Solids, and More Overview 3
National Education Reform Documents 5
Chinese Proverb 7

Lines, Line Segments, Points
Looking at Lines 9
Standing in Line 19
The Art of Geometry 25

Rays and Angles
Point the Ray 29
Angle Hunt 37
Angle Aerobics 41

2-D Shapes
Sorting Shapes 47
Sets of Shapes 55
Shape Shifters 57
Jump Rope Geometry 61
Polygon or Non? 75
Pondering Polygons 81
Tasty Triangles 93
Picking Out Shapes 101
Creating Congruence 107
Geometric Garden 115
Tangram Tinkerings 123
Cover Alls: Combining Tangrams 129
Animals Take Shape 135
Chart the Parts 149
Geometric Gallery 155
Shapes All Around Us 161

Circles
Radius Roundup 169
Delivering Diameters 175
All Around Circles 181
Circle Concepts 187

Symmetry and Transformations
Suitcase Symmetry 191
Sticking to Symmetry 199
Slip Slidin' 209
Flippin' Frogs 217
Flipping for Transformations 225
Picturing Rotations 235
The Pattern Block Shuffle 241
The Transforms 251
Quilted Transformations 259
Transformation Identification 271

3-D Objects
Something About Solids 279
Shape Shadows 287
Sorting Out Solids 293
Solid Shape Relay 299
Construction Zone 305
Cool Castles 313
3-D Designs 319
Prisms and Pyramids 325
Get a Clue 337

Location/Coordinate Grid
Shape Town 341
Where Is It? 349
Coordinate Camping 357
Coordinate Chorus 365
Ship Shape 367

Playful Practice
Ge-O Game 377
Who Has…Geometry 383
Triple Treasure Trivia 391
Word Search 399
Geometry Jeopardy 405
Shape Draw 411

Appendix
Literature Connections 413

SHAPES, SOLIDS, AND MORE

Geometry is everywhere. It is found in the architecture of our buildings. It is found in nature, in art, and in the construction of our bridges, our automobiles, and every other manufactured object. Students need to recognize geometric ideas and apply them in their everyday lives. *Shapes, Solids, and More* includes hands-on activities that will help second and third grade students do just that.

This book is divided into eight sections, one of which is *Playful Practice*. Each section provides grade-level appropriate activities that integrate the language of geometry into the development of the geometric concept. Much of geometry is the learning of its language. Geometry is often confusing as it contains many multiple meaning words such as lines, rays, faces, among others. Students are also introduced to new and difficult terms like congruent, symmetric, transformations, etc. Throughout this book, the language has been imbedded in the activity, thereby providing sustained reinforcement.

Learning Goals

Lines, Line Segments, Points
Identify and compare straight, curved, parallel, and intersecting lines.
Draw parallel and intersecting lines to create geometric shapes.

Rays and Angles
Compare rays to lines and line segments.
Identify and model acute, obtuse, and right angles.

2-D Shapes
Sort shapes into polygons and non-polygons.
Build, identify, compare, and contrast two-dimensional geometric shapes.
Become familiar with equilateral, isosceles, and scalene triangles.
Discover that congruent means same size and shape.
Compose and decompose various shapes.

Circles
Identify and measure the radius, diameter, and circumference of a circle.
Understand the relationship between radius and diameter.

Symmetry and Transformations
Examine real-world objects and drawings to determine whether they are symmetric.
Find the lines of symmetry on several geometric shapes.
Investigate, recognize, and apply slides (translations), flips (reflections), and turns (rotations).

3-D Objects
Compare and contrast characteristics of geometric solids.
Identify faces, edges, and vertices.
Combine various geometric solids to make specific structures.
Build three-dimensional structures.

Location/Coordinate Grid
Use positional words to describe an object's location.
Describe paths using directional words.
Locate objects on a coordinate grid.

National Education Reform Documents

The AIMS Education Foundation is committed to remaining at the cutting edge of providing curricular materials that are user-friendly, educationally sound, developmentally appropriate, and aligned with the recommendations found in national education reform documents.

Project 2061 Benchmarks

The Nature of Mathematics
- *Circles, squares, triangles, and other shapes can be found in things in nature and in things that people build.*
- *Numbers and shapes can be used to tell about things.*
- *Numbers and shapes—and operations on them—help to describe and predict things about the world around us.*

The Nature of Technology
- *When trying to build something or to get something to work better, it usually helps to follow directions if there are any or to ask someone who has done it before for suggestions.*

The Designed World
- *Several steps are usually involved in making things.*

The Mathematical World
- *Shapes such as circles, squares, and triangles can be used to describe many things that can be seen.*
- *Many objects can be described in terms of simple plane figures and solids. Shapes can be compared in terms of concepts such as parallel and perpendicular, congruence and similarity, and symmetry. Symmetry can be found by reflection, turns, or slides.*

Common Themes
- *Some features of things may stay the same even when other features change. Some patterns look the same when they are shifted over, or turned, or reflected, or seen from different directions.*

Habits of Mind
- *Assemble, describe, take apart and reassemble constructions using interlocking blocks, erector sets, and the like.*
- *Use numerical data in describing and comparing objects and events.*

NCTM Standards 2000*

Algebra
- *Sort, classify, and order objects by size, number, and other properties*

Geometry
- *Recognize, name, build, draw, compare, and sort two- and three-dimensional shapes*
- *Describe attributes and parts of two- and three-dimensional shapes*
- *Investigate and predict the results of putting together and taking apart two- and three-dimensional shapes*
- *Describe, name, and interpret relative positions in space and apply ideas about relative position*
- *Describe, name, and interpret direction and distance in navigating space and apply ideas about direction and distance*
- *Find and name locations with simple relationships such as "near to" and in coordinate systems such as maps*
- *Recognize and apply slides, flips, and turns*
- *Recognize and create shapes that have symmetry*
- *Create mental images of geometric shapes using spatial memory and spatial visualization*
- *Recognize and represent shapes from different perspectives*
- *Relate ideas in geometry to ideas in number and measurement*
- *Recognize geometric shapes and structures in the environment and specify their location*
- *Identify, compare, and analyze attributes of two- and three-dimensional shapes and develop vocabulary to describe the attributes*
- *Classify two- and three-dimensional shapes according to their properties and develop definitions of classes of shapes such as triangles and pyramids*
- *Investigate, describe, and reason about the results of subdividing, combining, and transforming shapes*
- *Explore congruence and similarity*
- *Describe location and movement using common language and geometric vocabulary*
- *Make and use coordinate systems to specify locations and to describe paths*

SHAPES, SOLIDS, AND MORE

- *Find the distance between points along horizontal and vertical lines of a coordinate system*
- *Predict and describe the results of sliding, flipping, and turning two-dimensional shapes*
- *Identify and describe line and rotational symmetry in two- and three-dimensional shapes and designs*
- *Build and draw geometric objects*
- *Recognize geometric ideas and relationships and apply them to other disciplines and to problems that arise in the classroom or in everyday life*

Measurement
- *Understand how to measure using nonstandard and standard units*
- *Measure with multiple copies of units of the same size, such as paper clips laid end to end*
- *Use tools to measure*
- *Understand such attributes as length, area, weight, volume, and size of angle and select the appropriate type of unit for measuring each attribute*

Problem Solving
- *Build new mathematical knowledge through problem solving*
- *Solve problems that arise in mathematics and in other contexts*
- *Apply and adapt a variety of appropriate strategies to solve problems*

Communication
- *Organize and consolidate their mathematical thinking through communication*

* Reprinted with permission from *Principles and Standards for School Mathematics*, 2000 by the National Council of Teachers of Mathematics. All rights reserved.

I Hear and
I Forget,

I See and
I Remember,

I Do and
I Understand.
—Chinese Proverb

LOOKING AT LINES

Topic
Lines

Key Question
How can you tell the difference between parallel and intersecting lines?

Learning Goals
Students will:
- build and compare straight and curved lines;
- build and compare parallel lines and intersecting lines; and
- identify examples of straight, curved, parallel, and intersecting lines.

Guiding Document
NCTM Standards 2000*
- *Describe, name, and interpret direction and distance in navigating space and apply ideas about direction and distance*
- *Create mental images of geometric shapes using spatial memory and spatial visualization*
- *Organize and consolidate their mathematical thinking through communication*

Math
Geometry
    lines

Integrated Processes
Observing
Comparing and contrasting
Classifying
Communicating
Relating

Materials
For each student:
    a set of Wikki Stix® (see *Management 1*)
    21 Teddy Bear Counters (see *Management 3*)
    metric rulers
    two pieces of construction paper, 12" x 18"
        (see *Management 2*)
    string
    copy paper
    picture page
    red sticky dot

Background Information
Shapes are made of straight, curved, parallel, and intersecting lines. This lesson begins with a simple look at *lines/line segments* that are used in varying arrangements to construct shapes.

As the students observe, construct, and define the various lines in this lesson, spatial awareness will begin to develop. Communication skills become important as the students use their own words along with the acquisition of new vocabulary to describe their experiences throughout the lesson. This study of lines builds a needed foundation for the study of the attributes of polygons (many-sided shapes).

See *Line Facts* for more information. At the primary level, the term *line* is used to refer to lines, line segments, and curved lines.

Management
1. Each student will need a minimum of 10 Wikki Stix®. 1-800-869-4554 (or) www.wikkistix.com.
2. Laminate a 12" by 18" piece of construction paper for each student to use as a building mat. The other piece of construction paper will be used for drawing lines.
3. Each student will need one red, 10 blue, and 10 green Teddy Bear Counters.
4. Each student will need string or additional Wikki Stix® to use in *Part Three*, step *3*.
5. Throughout this activity it should be stressed that parallel lines never intersect, they are an equal distance apart. It should also be stressed that intersecting lines always have a point of intersection. Sometimes line segments must be continued in order to find that point of intersection.

Procedure
Part One—Looking at Curved Lines
1. Draw various lines on chart paper or on the board. (Be sure to draw lines that are straight, vertical, horizontal, curved, simple closed curved, parallel, and intersecting.) Ask the students to describe the lines you have drawn. Explain that they are going to learn something about lines today.
2. Distribute Wikki Stix® and a piece of construction paper to use as a building mat to each student.
3. Direct the students to use a Wikki Stix® to form a curved line on their mats. Ask them to compare the orientation of these lines to those of others in their group. Discuss how some lines curve more than others, some curve up and down, others curve around and around, etc.

SHAPES, SOLIDS, AND MORE   © 2009 AIMS Education Foundation

4. Give the students some exploration time building designs/illustrations on their mats with their Wikki Stix®. Ask them to use only curved lines. Have them share their creations with the class.

Part Two—Checking for Student Understanding
1. Give each student a set of 10 Teddy Bear Counters and a sheet of construction paper. Ask the students to draw a curved line. Have them place their counters on top of the drawn lines to form curved lines of Teddy Bear Counters.
2. Have them compare their lines to a classmate's. Ask them to explain the direction in which their lines are curving. For example, curving up and down, around and around, up and under, etc.

3. Ask them to describe whether or not their curved lines are very curved, a little curved, etc. [e.g., My line is curved more than yours.]
4. Give the students some exploration time to use the Teddy Bear Counters on their mats to build curved lines. Have them make comparisons with their classmates.

5. Ask the students to illustrate these lines on a sheet of copy paper. Have them describe their curved lines.

Part Three—Looking at Parallel Lines
1. Direct students to place one Wikki Stix® on their mat in a straight line. Ask them to compare the orientation of these lines to those of others in their group. Discuss how some of the lines go across (horizontal), others go up and down (vertical), and others slant. Note that these lines are all considered to be straight and not curved.
2. Have the students use a second Wikki Stix® to build a line next to the first straight line on their mats. Tell them to use a ruler to check that the two lines are an equal distance apart at the beginning of the lines, in the center of the lines, and at the end of the lines. Ask them to compare the orientation of the two lines with a classmate's. [Some go up and down, others go across, and others are slanted. Some are very close, others are farther apart.] Tell the students that when two lines are side by side

and are equally spaced along the entire length of the lines, they are *parallel*. Ask the students to build a second set of parallel lines. Direct some of the students to build sets of lines that are at least four centimeters apart, other students to build sets of lines that are only one centimeter apart, and others that are more than eight centimeters apart. Discuss how all these lines are parallel.
3. Have the students take a second look at the parallel lines on their mats. Ask them to extend these lines as far as they can within the space on which they are working. Have them use a piece of string or additional Wikki Stix® to extend the lines. Point out that because the lines are an equal distance apart, the lines will never touch or intersect.
4. Give the students some exploration time to build designs/illustrations on their mats using only parallel lines. Have them share their mats with the class.

Part Four—Checking for Student Understanding
1. Ask students to draw a set of parallel lines on their construction paper. Discuss what they needed to do to accomplish this task. [Measure between the lines to make sure they are the same distance apart at the beginning, the middle, and at the end.]
2. Give each student a set of 10 blue and 10 green Teddy Bear Counters. Direct the students to place two lines of Teddy Bear Counters on the set of parallel lines.
3. Give the students some exploration time to build sets of parallel lines using only Teddy Bear Counters (do not have them draw the lines first). Have the students compare and describe how their lines are alike and how they are different from their classmates. [some are farther apart than others, some go up and down, some go across, etc.]
4. Invite the students to illustrate these lines on their paper. Have them describe what they know about parallel lines.

Part Five—Looking at Intersecting Lines
1. Give each student a set of Wikki Stix® and a building mat. Ask the students to place two Wikki Stix® on their mats in straight lines. Ask them to arrange them so that one line crosses the other. Invite them to compare the placement of these lines to those of others in their group. Discuss how these lines can be placed up and down, across, or on a slant. Explain that when a line touches or crosses another line, that it is called an *intersecting line*. Direct the students to place a red sticky dot at the point of intersection.
2. Ask the students to place two Wikki Stix® next to each other that are not equal distance apart. Make sure they do not touch.

3. Ask the students to use a piece of string or additional Wikki Stix® to show how if these lines were continued, they would intersect. Direct them to move their red sticky dot to the point of intersection of these lines.
4. Lead the class in a discussion about the "invisible intersection" of these lines. Help them to understand that lines will at some point intersect if they are not placed an equal distance apart.
5. Give the students some exploration time to build designs on their mats using only intersecting lines. Have them share their mats with the class.

Part Six—Checking for Student Understanding
1. Give each student a set of 10 blue and 10 green Teddy Bear Counters. Ask the students to use these counters to form two intersecting lines on their mats.
2. Once the students have constructed blue and green lines of Teddy Bear Counters that intersect, give each student one red Teddy Bear Counter. Have them remove the bear at the point of intersection and direct them to place the red bear at this point.

3. Have the students compare and describe how their lines are alike and how they are different from their classmates. [some cross in the middle, some cross/meet at the top/bottom, etc.]
4. Ask the students to draw a set of intersecting lines on their paper. Have them color a red dot at the point of intersection of the two lines. Have them describe what they know about intersecting lines.

Part Seven—Review
Give each student the picture that is included in this activity. Ask them to locate parallel, intersecting, straight, and curved lines on the picture. Direct them to place a red crayon mark at the point where lines intersect. Ask them to circle the parallel lines. Have them discuss their discoveries with a partner.

Connecting Learning
1. Describe lines that you have seen. [some are straight, some are curved, parallel, intersecting, etc.]
2. What is important to know about a set of parallel lines? [They are the same distance apart along the entire length of the lines.]
3. How can you tell the difference between a set of parallel lines and lines that are not parallel?
4. What are some things you have seen that have parallel lines?
5. What are some things you have seen that have straight lines? ...curved lines?
6. Show the class a straight line in the classroom.
7. Show the class a curved line in the classroom.
8. Show the class a set of parallel lines in the classroom.
9. What is important to know about a set of intersecting lines? [They touch or cross another line.]
10. What are some things you have seen that have intersecting lines?
11. Show the class a set of intersecting lines in the classroom.
12. Describe how a set of intersecting lines is different from a set of parallel lines.
13. Describe an example of more than two intersecting lines. [spokes on a bicycle, lines on a dart board, etc.]
14. When I write the word "parallel" on the board, are there any letters that form parallel lines? What are they? How many parallel lines are there in the word "parallel"? Does anyone have a name that has a set of parallel lines in it? If so, come write your name on the board.
15. Look at the letters v, w, x, and y. What do you notice about the lines that make up these letters? [They are all intersecting.]

* Reprinted with permission from *Principles and Standards for School Mathematics,* 2000 by the National Council of Teachers of Mathematics. All rights reserved.

Line Facts

These definitions have been included for two reasons:
1. for the teacher's professional growth and background, and
2. to show how abstract these concepts are, and to suggest to the teacher that the young child will not be expected to grasp the full concept of a *line*.

A *line* has no thickness and extends forever in two directions. A line is represented with arrows indicating that the line continues in both directions.

A *line segment* is a subset of a line that contains two points of the line and all points between those two points.

A *ray* is a subset of a line that contains one point and all points on the line on one side of the point.

A *closed curve* is a curve that, when traced, has the same starting and stopping points and may cross itself at individual points.

A *simple closed curve* is a curve that does not cross itself when traced. Its starting and stopping points are the same.

A *polygonal curve* is a subset of the simple closed curves. It is made up of line segments.

SHAPES, SOLIDS, AND MORE © 2009 AIMS Education Foundation

LOOKING AT LINES

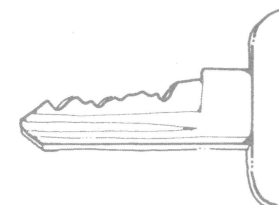

Key Question

What kind of lines can we make?

Learning Goals

- build and compare straight and curved lines;
- build and compare parallel lines and intersecting lines; and
- identify examples of straight, curved, parallel, and intersecting lines.

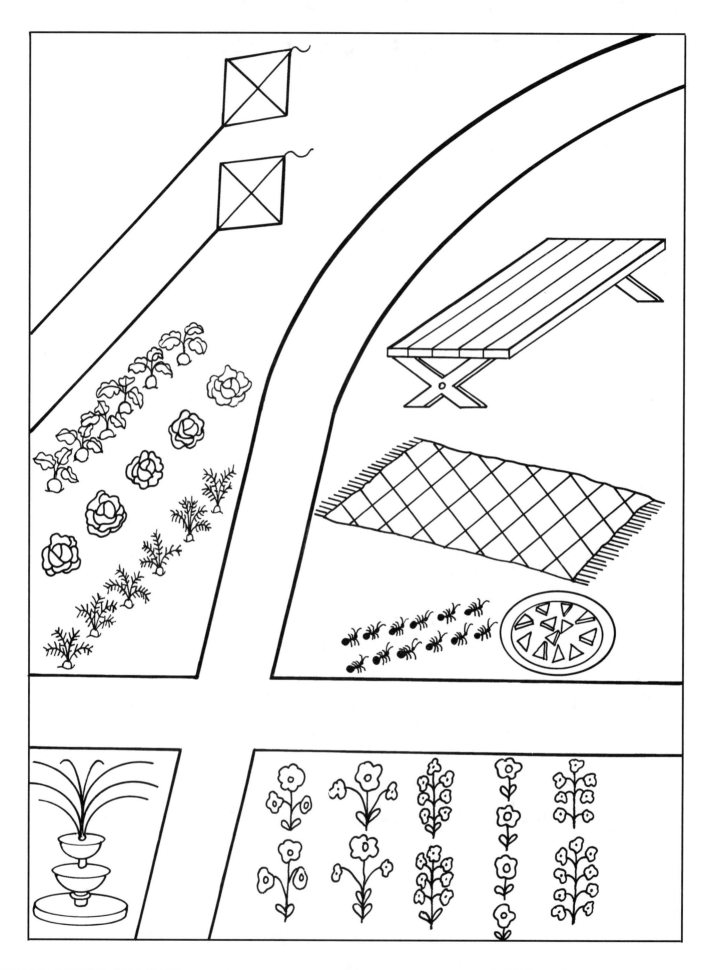

LOOKING AT LINES

Connecting Learning

1. Describe lines that you have seen.

2. What is important to know about a set of parallel lines?

3. How can you tell the difference between a set of parallel lines and lines that are not parallel?

4. What are some things you have seen that have parallel lines?

5. What are some things you have seen that have straight lines? ...curved lines?

6. Show the class a straight line in the classroom.

LOOKING AT LINES

Connecting Learning

7. Show the class a curved line in the classroom.

8. Show the class a set of parallel lines in the classroom.

9. What is important to know about a set of intersecting lines?

10. What are some things you have seen that have intersecting lines?

11. Show the class a set of intersecting lines in the classroom.

12. Describe how a set of intersecting lines is different from a set of parallel lines.

LOOKING AT LINES

Connecting Learning

13. Describe an example of more than two intersecting lines.

14. When I write the word "parallel" on the board, are there any letters that form parallel lines? What are they? How many parallel lines are there in the word "parallel"? Does anyone have a name that has a set of parallel lines in it? If so, come write your name on the board.

15. Look at the letters v, w, x, and y. What do you notice about the lines that make up these letters?

Standing in Line

Topic
Lines

Key Question
How are straight and curved lines different?

Learning Goals
Students will:
- build and compare straight and curved lines; and
- build and compare parallel and intersecting lines.

Guiding Document
NCTM Standards 2000*
- *Describe, name, and interpret direction and distance in navigating space and apply ideas about direction and distance*
- *Organize and consolidate their mathematical thinking through communication*

Math
Geometry
    lines

Integrated Processes
Observing
Comparing and contrasting
Communicating
Relating
Applying

Materials
For each group of students:
    measuring tool (see *Management 1*)
    one piece of red construction paper
        (see *Management 2*)
    2 rolls of crepe paper streamers or adding machine tape (see *Management 3*)

For each student:
    one piece of construction paper, 12" x 18"

Background Information
   To continue the study of lines, this lesson is designed to meet the needs of the bodily kinesthetic learners. The students are asked to place themselves in lines depicting curved, straight, parallel, and intersecting lines. This lesson is also used to reinforce their understanding of the attributes of these lines.
   Once the students have the foundation of knowledge of these types of lines, they will better be able to explore the attributes of shapes.

Management
1. Provide a measuring tool, such as a measuring tape, for each group of students.
2. Cut a large circle from the red paper. Use the circle to mark the point of intersection of the two lines.
3. Provide crepe paper streamers or strips of adding machine tape to use to mark the lines the children form.
4. Locate a large area (blacktop, cafeteria, all-purpose room) where students can construct "student" curved, straight, parallel, and intersecting lines.

Procedure
Part One—Curved Lines
1. Take the class to a large area. Ask them to form a curved line. Give the first student in line a roll of crepe paper streamer (or adding machine tape). Tell the students that they are going to pass this roll from one student to the next, holding onto the strip as they unroll it. Tell them all to hold it in the same hand, left or right.
2. When all students have the paper, tell them to bend over and place it beside their foot, the one on the same side as the hand that is holding the paper. Invite the students to step away from the paper in order to observe and discuss the curve. Ask the students to describe the line (e.g., a small curve, a large curve, curve up, curve down, curve up and down, curve around, etc.).
3. Divide the class in half. Have each group form separate curved lines. Using two rolls of paper, ask them to form the lines in the same manner as above and place them on the ground.
4. Gather the entire class around each set of curved lines and discuss what the curved lines look like (e.g., one is a very small curve, another has a very large curve, one curves up and down, another curves around and around, etc.).
5. Ask the students if they can think of anything they have seen that would be made of curved lines. [a walkway through the park, a road through the mountains, etc.]
6. Take the students to the playground or classroom and have them create curved roads in the sandbox using blocks or other school materials.

Part Two—Straight Lines and Curved Lines
1. As the students line up for lunch or a trip to the library, etc., have them form a straight line. Use a paper streamer to duplicate this line on the floor next to the "student" line.

SHAPES, SOLIDS, AND MORE   © 2009 AIMS Education Foundation

2. Next to this line, have the students line up again, forming a curved line. Duplicate this line using a paper streamer next to the straight line.
3. Have the students step back and observe the lines. Ask them to discuss how the lines are different.
4. Have the students return to their line to get ready to go. On the way to lunch, the library, etc., ask the students to point out straight lines and curved lines they observe around the school.

Part Three—Parallel Lines
1. Take the students outdoors and ask half the class to form a straight line. Ask the other half of the class to form a line that is parallel to the first line. Using two rolls of paper, have the students mark their two lines on the ground. Have them move away from the lines in order to look at them and to compare them. Help them to measure to make sure the two lines are the same distance apart at the beginning, the middle, and the end of the lines. If necessary, adjust the paper lines so they are parallel.

2. Challenge the class to construct other sets of parallel lines in a different arrangements—vertical, horizontal, diagonal. Have them duplicate the lines using paper streamers.
3. Gather the entire class around each set of lines and discuss how each set is parallel even though they are not in the same arrangement.
4. Ask the children if they can think of anything they have seen that would be made of parallel lines. [railroad tracks, race car toy tracks, sidewalks on either side of the street, etc.]

Part Four—Intersecting Lines
1. Ask the class to line up in a set of parallel lines. Discuss what they would need to do to their lines to change them to a set of intersecting lines.
2. Have them try out their suggestions. Direct them to mark each configuration with paper streamers on the ground.

3. Discuss how these lines are called *intersecting lines* because they run into each other, or cross over each other. Give the student who is standing at the point of intersection a red circle to hold up.

4. Divide the class in half and have the two groups form separate sets of intersecting lines, marking them with paper streamers. Give each group a red circle to place at the intersection of their lines.

5. Gather the class around each set of intersecting lines to discuss how the orientations of the lines are alike and how they are different.
6. Direct the class to line up to form two straight lines. Move these lines side by side so that they are not parallel but do not intersect. Ask the students to describe these lines using their new vocabulary. (Most students will call these parallel lines.) Discuss how lines need to be thought of as continuing on past the lines they actually see. Demonstrate this by using the rolls of paper to mark the student lines and beyond until the two lines intersect. Review how these lines are not parallel because they meet at some point when continued in this manner (the invisible intersection). Place a red circle to denote the intersection.
7. Invite the class to form lines that will intersect once a paper streamer line is continued. Ask classmates to check the lines for the "invisible intersection" using paper streamers. Discuss each arrangement with the whole class.

Part Five—Review
1. Take the students outside on a "line search" to locate examples of parallel, intersecting, straight, and curved lines.
2. When students find examples of the different types of lines, invite them to do a shout out of the example. At this time, the class is to turn its attention to the example and determine if the caller is correct. Inform students that they cannot call out lines that have already been called out.
3. Return to the classroom. Distribute a 12" x 18" sheet of paper to each student. Have them make a drawing that contains straight lines, curved lines, some parallel lines, and some intersecting lines.
4. Ask students to trade their pictures with a partner. Invite the partners to find and identify the various types of lines on their classmate's illustration.
5. Display the pictures for all to see. Tell students that they are welcome to look at the pictures to find the types of lines when they have free time.

Connecting Learning
1. How are parallel lines different from intersecting lines?
2. Are all straight lines intersecting lines? Explain your answer.
3. Can parallel lines be intersecting lines? Explain your answer.
4. Which type of lines did we find the most examples of when you went outdoors?
5. Were you able to find all the types of lines in your partner's drawing? Explain.
6. What kind of lines are found in the letter H? [parallel and intersecting lines]
7. What kinds of lines do you use when you write your name?

* Reprinted with permission from *Principles and Standards for School Mathematics,* 2000 by the National Council of Teachers of Mathematics. All rights reserved.

Standing in Line

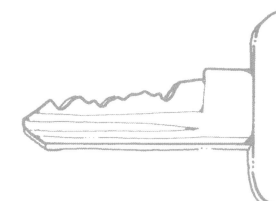

Key Question

How are straight and curved lines different?

Learning Goals

Students will:

- build and compare straight and curved lines; and
- build and compare parallel and intersecting lines.

Standing in Line

Connecting Learning

1. How are parallel lines different from intersecting lines?

2. Are all straight lines intersecting lines? Explain your answer.

3. Can parallel lines be intersecting lines? Explain your answer.

4. Which type of lines did we find the most examples of when you went outdoors?

5. Were you able to find all the types of lines in your partner's drawing? Explain.

6. What kind of lines are found in the letter H?

7. What kinds of lines do you use when you write your name?

The Art of Geometry

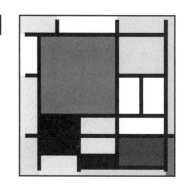

Topic
Lines

Key Question
How is geometry related to art?

Learning Goals
Students will:
- draw a series of parallel and intersecting lines to create geometric shapes, and
- reproduce the art style of Piet Mondrian.

Guiding Document
NCTM Standard 2000*
- *Recognize geometric shapes and structures in the environment and specify their location*

Math
Geometry
   lines
      parallel, intersecting

Integrated Processes
Observing
Comparing and contrasting
Applying

Materials
White paper
Red, yellow, and blue paints, markers, or crayons
Black marker
Ruler
Pencil

Background Information
    Geometric shapes are everywhere, from the foods we eat to the homes we live in to the art we enjoy. We can gain an awareness and appreciation of the geometry in our world through observation. The more focused our observations become, the more details we notice.
    In this activity, students will be exposed to an artist, Piet Mondrain, and the grid-based painting style he developed. Piet Mondrian was a Dutch painter born in 1872. At one time, he painted realistic landscapes, but the more he painted, the more his style began to change. In 1919 he began to create abstract images. The new style developed as Mondrian looked closer at the trees, buildings, and vases he painted. Soon he saw their basic shapes and colors. He began to look at nature through squinted eyes and all the details started to disappear. He began to see only shapes and color, no real objects. Eventually, Mondrian's style consisted of geometric lines, shapes, and primary colors. His paintings now hang in the Museum of Modern Art in New York City and the Philadelphia Museum of Art.

Management
1. Samples of Piet Mondrian's art can be found at the following websites:
   http://www.ibiblio.org/wm/paint/auth/mondrian/
   http://en.wikipedia.org/wiki/Piet_Mondrian

Procedure
1. Start the lesson by asking the class how they think art and geometry might be related. Give students background on Piet Mondrian and the painting style he developed. (See *Background Information*.)
2. Display samples of Mondrian's work. Explain to the class that they are going to reproduce Piet Mondrian's style using a series of parallel and intersecting lines.
3. Give each student a white piece of paper, ruler, and paints, markers, or crayons in red, yellow, and blue.
4. Instruct the class to divide their papers by drawing four horizontal lines from one end of the paper to the other. Ask them to draw three vertical lines, making sure the lines go from one end of the paper to the other. When the lines have been drawn, have them use a black marker to darken them. Suggest that they make some lines thick and some thin. When students have completed their lines, tell them to choose just a few rectangular spaces on their papers to fill in with the primary colors—red, yellow, and blue. Encourage students to leave some white space. Have students sign their work and use the art for a wall display.

5. When all artwork is complete, end with a discussion about how art and geometry are related. Question students about the shapes and lines involved in their pictures.

Connecting Learning
1. Where in the real world do we see geometric shapes and lines?
2. How are art and geometry related?
3. What types of lines did you use in your artwork?
4. Identify some parallel lines in one of the art pieces. ...intersecting lines.
5. What shapes did you fill with color?
6. How might you find out about other artists that use geometric shapes and lines in their art?

* Reprinted with permission from *Principles and Standards for School Mathematics*, 2000 by the National Council of Teachers of Mathematics. All rights reserved.

The Art of Geometry

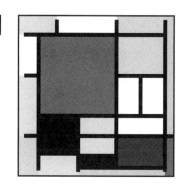

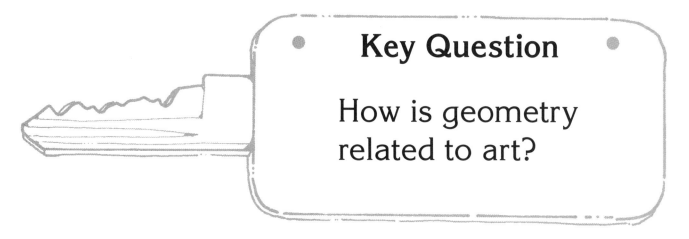

Key Question

How is geometry related to art?

Learning Goals

Students will:

- draw a series of parallel and intersecting lines to create geometric shapes, and
- reproduce the art style of Piet Mondrian.

The Art of Geometry

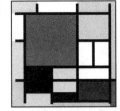

Connecting Learning

1. Where in the real world do we see geometric shapes and lines?

2. How are art and geometry related?

3. What types of lines did you use in your artwork?

4. Identify some parallel lines in one of the art pieces. ...intersecting lines.

5. What shapes did you fill with color?

6. How might you find out about other artists that use geometric shapes and lines in their art?

Point the Ray

Topics
Lines, line segments, rays

Key Question
How are lines, line segments, and rays different?

Learning Goal
Students will model and compare parallel and intersecting lines, line segments, and rays.

Guiding Document
NCTM Standards 2000*
- *Identify, compare, and analyze attributes of two- and three-dimensional shapes and develop vocabulary to describe the attributes*
- *Recognize geometric ideas and relationships and apply them to other disciplines and to problems that arise in the classroom or in everyday life*

Math
Geometry
  lines
  line segments
  rays

Integrated Processes
Observing
Comparing and contrasting
Applying

Materials
A Look at Lines Journal (see *Management 1*)
String (see *Management 2*)
Black construction paper (see *Management 4*)

Background Information
   A *line* can be described as infinitely long, perfectly straight, and containing an infinite number of points. *Line segments* and *rays* can be defined in terms of a line. A *line segment* is a part of a line with a beginning point and an ending point. Mathematicians call those points *endpoints*. A *line segment* has a finite length. A *ray* is a part of a line with a beginning point but no ending point. A ray goes on infinitely in one direction. A model for a ray usually has an arrowhead at one end. A *ray* starts at a given point and goes off in a specific direction to infinity. The point where the r*ay* starts is called the *endpoint*.

Management
1. Each student needs a copy of the *A Look at Lines Journal* for *Part One*. To make a journal, copy the journal pages and cut them in half. Stack them in order and staple along the left edge. Each journal will have five pages, including the cover. The second page of the journal (This is a line segment because; This is a ray because) is printed twice on the same page to reduce the number of copies that need to be made.
2. Divide the class into groups of two for *Part Two* of this lesson. Each group will need a two-meter piece of string.
3. A large area for students to stretch out is needed for *Part Two* of this lesson.
4. Prior to *Part Two* of this lesson, give each pair of students a 12" x 18" piece of black construction paper. Instruct them to fold it in half vertically and 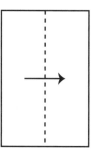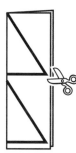 draw two large triangles on the fold to represent arrows. Have students cut out the arrows.

Procedure
Part One: Introducing Terms
1. Draw a point on the board and define it as a *point*. Discuss the fact that you could make a line by drawing millions of points. Talk about the fact that even though a line truly contains an infinite number of points, we do not draw a line using points because it would take too much time.
2. Draw a line on the board. (There should be an arrow on each end). Ask the class to identify what you drew. (They will generally say a line but not know why it is a line.) Explain that they are correct if they said that it was a line. Question them about characteristics of a line. [It goes on forever in both directions, it contains an infinite number of points, etc.]
3. Distribute the *A Look at Lines Journal*. Ask students to turn to the first page and draw a point and a line in the correct boxes. Have students describe in their own words what points and lines are.

SHAPES, SOLIDS, AND MORE

4. When students have finished writing about the point and line, draw their attention back to the line on the board. Ask students what would happen if you took a piece of the line right out of the center. Erase a section of the line and draw a smaller line below the first line. Discuss how this piece or segment of the line won't go on forever, so it should not have arrows at each end but points to show that it stops. Explain that mathematicians call these stopping points *endpoints*.

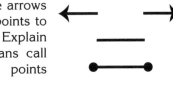

5. Draw a ray on the board. (There should be an arrow on one end and an endpoint on the other.) Question the class about what you have drawn. Ask them what the arrow on one end and the point on the other mean.
6. Ask students to turn to the next page in the journal and draw and write about the line segment and ray.
7. Draw a pair of parallel lines on the board. Ask the students what they notice about the two lines. Discuss the fact that they go on forever in both directions and because they are equal distances apart, they are called parallel lines. Identify several sets of parallel lines in the classroom. [Two lines on a brick wall, the top and bottom of the board, etc.]
8. Draw a pair of intersecting lines on the board. Ask students to compare the new lines to the parallel lines previously discussed. Explain to the students that these lines are called intersecting lines because they meet or intersect each other.
9. Draw two lines on the board that are not touching but when extended will meet, or intersect. Tell the students that these two lines are also intersecting lines. Invite students to explain why they think the two lines, which are clearly not touching, would be considered interesting lines. [By definition, lines go on forever in both directions, therefore they will eventually meet.]

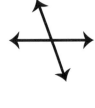

10. Allow time for the students to draw and write about parallel and intersecting lines.
11. Draw one last set of intersecting lines on the board—perpendicular to each other. Ask students to identify the type of lines. Explain that these are a special type of intersecting lines because they form a right angle at the intersection. Introduce the term *perpendicular*. Find examples of perpendicular lines in the classroom. Remind students that perpendicular lines can also be separated as long as they will form a right angle when extended.
12. When students have completed their journal entries by writing about perpendicular lines, end with a discussion about what they learned by looking closer at lines.

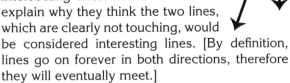

Part Two: Modeling Lines
1. Review points, line segments, rays, lines, intersecting, and parallel lines.
2. Move the class to a large open area. Give each pair of students a two-meter piece of string and two arrows.
3. Explain to the class that they will be using themselves, the string, and the arrows to model lines, rays, etc.
4. Point to a pair of students and say, "Show me a ray." While two students are modeling a ray by holding one end of the string in a tight fist and placing the arrow at one end of the string, ask another group if it truly is a ray and what makes it a ray. [It has an endpoint and extends forever in the opposite direction.]
5. Repeat this procedure several times allowing different groups to demonstrate a line, a line segment, and a ray.
6. When the students have demonstrated an understanding of the line, line segment, and ray concepts, combine the students into groups of four and ask the four students to use themselves and the two strings to demonstrate parallel and intersecting lines.
7. End with a time of discussion where students talk about similarities and differences between the terms learned and the value of making the string models.

Connecting Learning
1. Describe a point. …a line segment. …a ray.
2. How is a line segment different from a line? [It has two endpoints. A line goes on forever in each direction.] How can we tell a line segment just by looking at it? [It doesn't have arrows on its ends.]
3. Where do you think the term *line segment* comes from?
4. How could we use our bodies to model a ray? …a line? …a line segment?
5. How did modeling the lines, rays, and line segments help you gain a better understanding of them?

* Reprinted with permission from *Principles and Standards for School Mathematics*, 2000 by the National Council of Teachers of Mathematics. All rights reserved.

Point the Ray

Key Question

How are lines, line segments, and rays different?

Learning Goal

Students will:

model and compare parallel and intersecting lines, line segments, and rays.

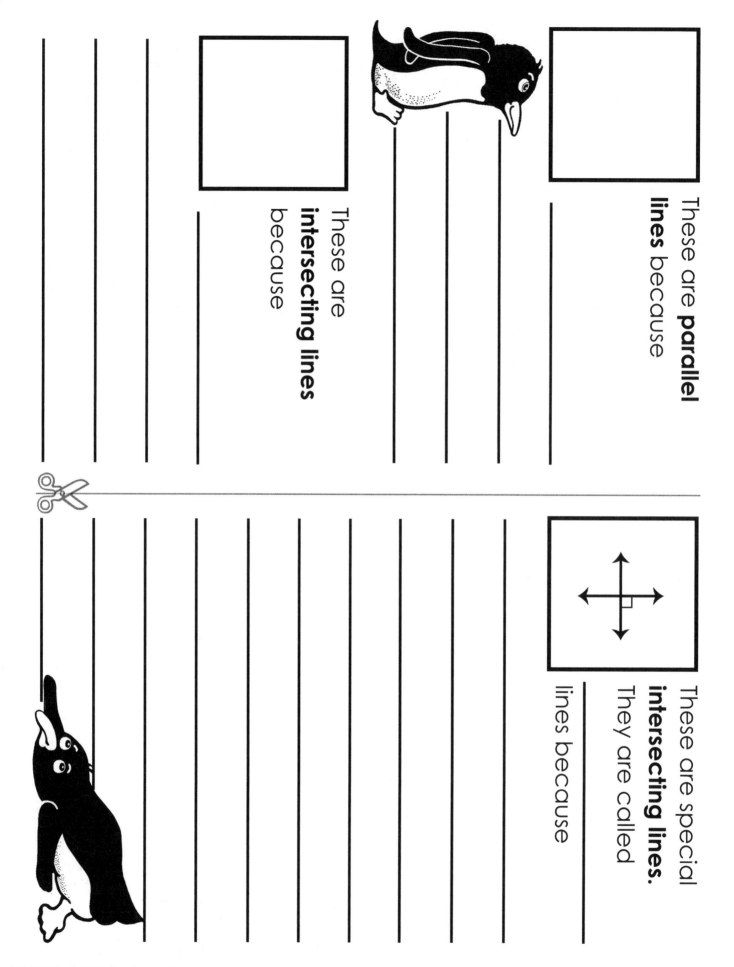

Connecting Learning

1. Describe a point. …a line segment …a ray.

2. How is a line segment different from a line? How can we tell a line segment just by looking at it?

3. Where do you think the term *line segment* comes from?

4. How could we use our bodies to model a ray? …a line? …a line segment?

5. How did modeling the lines, rays, and line segments help you gain a better understanding of them?

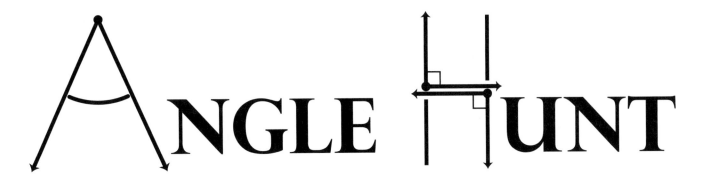

Topic
Angles

Key Question
What happens when two rays are connected at the same endpoint?

Learning Goals
Students will:
- identify angles as acute, obtuse, or right; and
- find real-world examples of each.

Guiding Documents
Project 2061 Benchmark
- *Numbers and shapes—and operations on them—help to describe and predict things about the world around us.*

NCTM Standards 2000*
- *Identify, compare, and analyze attributes of two- and three-dimensional shapes and develop vocabulary to describe the attributes*
- *Recognize geometric ideas and relationships and apply them to other disciplines and to problems that arise in the classroom or in everyday life*

Math
Geometry
    angles
        right, acute, obtuse

Integrated Processes
Observing
Comparing and contrasting
Applying

Materials
Paper with square corners
Black construction paper, one sheet per student
Toothpicks, six per student
Glue
White crayons

Background Information
The study of angles logically follows the study of lines, line segments, and rays because an angle is made up of two rays with a common endpoint (vertex). The measure of an angle can be taken by subdividing it into uniform wedge-shaped units (degrees). Just as the centimeter is a repeated unit used to measure length, the wedge is the repeated unit in angle measurement. For second and third grade students, it is not important that they can use a protractor to measure the size of an angle to the nearest degree. It is more important for them to gain a general understanding of how angles are classified. A benchmark in the world of angles is the *right angle* (90 degrees). To connect this new vocabulary to previously learned information, a right angle can be demonstrated by showing students a set of perpendicular intersecting lines. Often to make it easier for young children to remember what a right angle looks like, we say that a right angle makes a perfect capital L, or that the corner of a paper can be a "right-angle checker."

When students have a good understanding of right angles and can easily identify them in the real word, two other types of angles should be introduced. The new terms address angles that are larger than a right angle (*obtuse*) and angles that are smaller than a right angle (*acute*).

In this activity, students will be introduced to three types of angles. They will make models of the angles with toothpicks and will locate examples of the angles in the classroom.

Management
1. It is suggested that you use a colorful piece of paper for your angle checker if you are using it on a white board. This will allow students to better see the edges.

Procedure
1. Ask students if they remember what lines, line segments, and rays are.
2. Lead the class in a discussion that identifies an angle as two rays that are connected. Explain that angles come in many sizes and that angles are measured in degrees.

SHAPES, SOLIDS, AND MORE   37   © 2009 AIMS Education Foundation

3. Ask the students if they have ever heard anyone say, "I can do a 360 on my skateboard." Or "I can do a 180 on my bike." Have students share what they think these statements mean.
4. Draw a circle on the board. Explain that 360 is the number of degrees in a complete circle, so if someone can do a 360 on their skateboard, he/she can turn a complete circle. If the skateboarder is doing a 180, it means a turn half way around. Discuss how 180-degrees is half of the 360-degree circle. Draw a horizontal line across the circle to demonstrate.
5. Draw a vertical line that dissects the 180-degree line, forming perpendicular lines. Discuss how the lines divide the circle into four equal angles. Draw students' attention to the point of intersection and question them about what they think the size of the four angles would be. Explain that as far as angles go, one of the most important angles to learn about is the kind that they see at the intersection of the perpendicular lines. Introduce the term *right angle*. Explain that it is a 90-degree angle and that if it helps them remember what a right angle looks like, they can think of it as forming a capital L that can be turned in any direction.
6. Ask the class to show you with their fingers what a right angle looks like. Explain that they can use the square corner of a piece of paper as a right-angle checker. Draw an angle on the board and place the corner of the paper into the corner of the angle. Ask students if the edges of the paper fit perfectly. Explain that if they do, then it is a right angle.
7. Give each student a piece of black construction paper, glue, and six toothpicks. Assist them in dividing the paper into three sections. Ask them to use two of the toothpicks to show a right angle. When you have checked to see that they truly modeled a right angle, instruct the students to glue them down and, using a white crayon, label them "right angle."

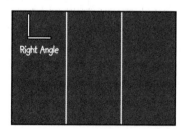

8. Give students each a right-angle checker and ask them to find right angles in the classroom and record what they find under the toothpick right-angle model.

9. Bring the class back for a time of sharing the right angles that they found. In some cases, ask the students to prove that the object they have identified has a right angle by using the right-angle checker.
10. Question the students about whether they found angles that were larger than a right angle. Draw an obtuse angle (greater than 90 degrees) on the board. Introduce the term *obtuse*. Ask the class how their right-angle checker might help them identify obtuse angles.
11. Have the students position the second set of toothpicks to form an obtuse angle in the second section of their papers. Check that they are correct, and instruct them to glue the toothpicks down and label their models. Allow time for students to locate several obtuse angles in the classroom and record them below their models.
12. Repeat this procedure, introducing *acute* angles (less than 90 degrees).
13. End with a time of sharing the real-world examples of the three types of angles.

Connecting Learning
1. What does it mean when someone does a 360 on his/her skateboard?
2. Show me an acute angle. ...right. ...obtuse.
3. How could you use your hands to model a right angle?
4. What is a right-angle checker, and how do you use it?
5. How can the right-angle checker be used to determine if an angle is obtuse or acute?

* Reprinted with permission from *Principles and Standards for School Mathematics*, 2000 by the National Council of Teachers of Mathematics. All rights reserved.

Angle Hunt

Key Question

What happens when two rays are connected at the same endpoint?

Learning Goals

Students will:

- identify angles as acute, obtuse, or right; and
- find real-world examples of each.

ANGLE HUNT

Connecting Learning

1. What does it mean when someone does a 360 on his/her skateboard?

2. Show me an acute angle. ...right ...obtuse.

3. How could you use your hands to model a right angle?

4. What is a right-angle checker, and how do you use it?

5. How can the right-angle checker be used to determine if an angle is obtuse or acute?

Angle Aerobics

Topic
Angles

Key Question
How can we exercise and model angles at the same time?

Learning Goals
Students will:
- identify acute, obtuse, or right angles; and
- model each angle type with their bodies.

Guiding Documents
Project 2061 Benchmark
- *Numbers and shapes—and operations on them—help to describe and predict things about the world around us.*

*NCTM Standards 2000**
- *Identify, compare, and analyze attributes of two- and three-dimensional shapes and develop vocabulary to describe the attributes*
- *Recognize geometric ideas and relationships and apply them to other disciplines and to problems that arise in the classroom or in everyday life*

Math
Geometry
    angles
        right, acute, obtuse

Integrated Processes
Observing
Comparing and contrasting
Communicating
Applying

Materials
Large open area
Wood slats (see *Management 2*)

Background Information
In this activity the three types of angles, *right, acute, and obtuse* introduced in the activity *Angle Hunt* will be reviewed. Students will then model them while doing toe touches and sit-ups.

Management
1. A large open area will be needed for students to do the angle aerobics.
2. If you have students who are limited in their mobility, simply drill a hole in two wood slats or craft sticks. Attach the sticks together with a paper fastener to form the two rays of an angle.
3. If you are uncomfortable doing sit-ups and toe touches, ask for a student volunteer to model the angle exercise moves.
4. Copy station cards on card stock and laminate for extended use.

Procedure
Part One
1. Review right, acute, and obtuse angles.
2. Have students stand by their desks with their hands at their sides. Ask them to show you what a right angle would look like, using their arms as the rays of the angle. Do the same for acute and obtuse.
3. Move the students to a large open area. Explain to them that they will now get to participate in some angle aerobics. Tell the class that you will first demonstrate the moves then will call out the moves for them to do.
4. Position yourself where everyone can see you. Turn sideways, put your arms up straight into the air, and bend at the waist with both arms out in front to form a right angle with your body. Ask the students what angle you are demonstrating. Have the students make the same angle with their bodies.
5. Have students observe you again as you model an acute angle. Stand up straight with arms over your head then bend until your arms are about a foot away from your feet. Have students identify the acute angle. Ask the students to make an acute angle with their bodies.
6. Repeat procedure five, this time holding your outstretched arms above your head and bending slightly at the waist to demonstrate an obtuse angle.
7. When students are familiar with the correct positions for right, obtuse, and acute angles, ask them to stand up straight and follow your directions. Begin calling out *right, obtuse, acute* in random order, allowing the students to get some good exercise while practicing three important angles.

SHAPES, SOLIDS, AND MORE   © 2009 AIMS Education Foundation

8. Explain that they will now do sit-up angle aerobics. Have students position themselves flat on the floor. Ask them to leave their legs straight out in front of them and to rise to a sitting position. Have students identify the angle they are forming with their bodies. [right]
9. Ask students to lean way forward to form acute angles, and lean slightly backward to form obtuse angles.
10. When students are clear about the position for each angle, again in random order, call out *acute, right,* and *obtuse.*
11. End with a discussion about other exercises that might involve making angles.

Part Two
1. Place the six station cards around the room. Divide the class into six groups, assigning each group a starting station. Explain the rotation pattern to students so they know the order they are to use to progress through all six stations.
2. Allow time for students to perform the aerobics at each station.
3. End with a class discussion about the types of angles they formed.

Connecting Learning
1. How can we exercise and model angles at the same time?
2. If I am nearly touching my toes, what type of angle am I modeling?
3. How could I model a right angle with my body?
4. What is an angle?
5. What other exercises involve making angles?

* Reprinted with permission from *Principles and Standards for School Mathematics,* 2000 by the National Council of Teachers of Mathematics. All rights reserved.

Angle Aerobics

Key Question

How can we exercise and model angles at the same time?

Learning Goals

Students will:

- identify acute, obtuse or right angles; and
- model each angle type with their bodies.

Sitting Right Angle
- Sit on the floor with your legs outstretched. Now lie back.
- Do 10 sit-ups that make your body form a right angle.
- When finished, remain sitting for 30 seconds to rest your body.
- Stand up and wait until you are told to go to the next station.

Sitting Obtuse Angle
- Sit on the floor with your legs outstretched. Now lie back.
- Do 10 sit-ups that make your body form an obtuse angle.
- When finished, remain sitting for 30 seconds to rest your body.
- Stand up and wait until you are told to go to the next station.

Sitting Acute Angle
- Sit on the floor with your legs outstretched. Now lie back.
- Do 10 sit-ups that make your body form an acute angle.
- When finished, remain sitting for 30 seconds to rest your body.
- Stand up and wait until you are told to go to the next station.

Standing Right Angle
- Stand with your arms above your head.
- Lean forward so that your body forms a right angle.
- Do 10 right angle bends.
- Wait until you are told to go to the next station.

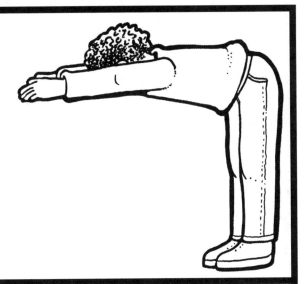

Standing Obtuse Angle
- Stand with your arms above your head.
- Lean forward so that your body forms an obtuse angle.
- Do 10 obtuse angle bends.
- Wait until you are told to go to the next station.

Standing Acute Angle
- Stand with your arms above your head.
- Lean forward so that your body forms an acute angle.
- Do 10 acute angle bends.
- Wait until you are told to go to the next station.

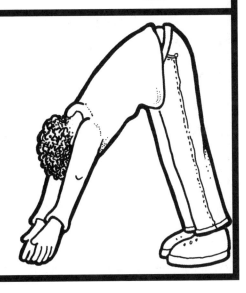

Angle Aerobics

Connecting Learning

1. How can we exercise and model angles at the same time?

2. If I am nearly touching my toes, what type of angle am I modeling?

3. How could I model a right angle with my body?

4. What is an angle?

5. What other exercises involve making angles?

Sorting Shapes

Topic
2-D shapes

Key Question
What rule was used to sort the shapes, and where do the remaining shapes belong, based on that rule?

Learning Goals
Students will:
- identify the rule used to sort a set of shapes, and
- place additional shapes where they belong based on that rule.

Guiding Document
NCTM Standards 2000
- *Sort, classify, and order objects by size, number, and other properties*
- *Recognize, name, build, draw, compare, and sort two- and three-dimensional shapes*

Math
Geometry
    properties of shapes
Sorting
    Venn diagrams

Integrated Processes
Observing
Comparing and contrasting
Identifying

Materials
Card sets
Crayons (red, green, blue)
Scissors
Grouping circles, optional

Background Information
    Sorting a variety of two-dimensional shapes can help students pay close attention to the characteristics of those shapes. It also helps reinforce that not all shapes of the same type look the same—not all triangles are equilateral, not all pentagons or hexagons are regular, etc. This activity provides an open-ended sorting experience than can be adjusted to the needs and abilities of your students.

Management
1. Copy the sets of shape cards onto card stock. Each pair of students needs one set of cards.
2. Make a set of cards for yourself. You will need to be able to attach these shapes to the board and move them around. If possible, use magnets. Otherwise, you will need to stick a loop of tape on the back of each card.
3. To make the activity simpler, eliminate certain shapes from the cards provided.
4. If available, attach grouping circles to the board to make the Venn diagrams. Otherwise, draw the circles by hand. Grouping circles are available from AIMS (#4621).

Procedure
1. Distribute the materials to each pair of students. Have students color the shapes as indicated and then cut out the cards.
2. Explain that you will be sorting some of the shapes into two groups on the board. Students will have to work with their partners to try and figure out the rule used for sorting and put all of the remaining cards where they belong.
3. Place a single grouping circle on the board and draw a large rectangle around it to indicate the two areas for sorting. Discuss with the class that the shapes you place in the circle will all have the same characteristic. The shapes that do not have that characteristic will go inside the rectangle but outside the circle.
4. Inside the circle, put the trapezoid, the square, and the rectangle. Outside the circle but inside the rectangle, put a triangle, a pentagon, a hexagon, and an octagon.
5. Instruct students to sort their shape cards on their desks like you have them sorted on the board. Challenge them to place the remaining 11 cards where they belong.
6. When pairs have finished their sorts, hold up one of the cards that has not been sorted and ask a student where it belongs. [Do not allow them to state why.] If he/she is correct, place the card in the appropriate location and move on to the next card. If the student is incorrect say, "I don't think you know the rule yet," and ask another student to identify where the card goes.

SHAPES, SOLIDS, AND MORE   © 2009 AIMS Education Foundation

7. Continue until all the cards have been placed, then ask the class to identify the rule you used for sorting. [quadrilaterals or not quadrilaterals]
8. Repeat this process with a variety of different sorts into two or three non-intersecting groups that focus on shape and/or color. For example, odd sides or even sides; red shapes, green shapes, or blue shapes; shapes with more than four sides or fewer than four sides; shapes with right angles, shapes without right angles, etc.
9. If appropriate for your students, create a two-circle Venn diagram on the board and do several sorts with intersecting sets, each time having them place the remaining cards where they belong. For example, red shapes and shapes with more than three sides; blue shapes and quadrilaterals, etc.

Connecting Learning
1. How were you able to determine the rule used for sorting the shapes?
2. Which sorts were the easiest to identify? ...the most difficult?
3. What other ways can you think of to sort the shapes?
4. What does it mean when there are two intersecting circles on a Venn diagram? [The shapes in the intersecting portion have characteristics of both attributes.]

Extension
Have pairs of students use their cards to make their own sorts, then challenge other pairs of students to determine the rule used for sorting.

* Reprinted with permission from *Principles and Standards for School Mathematics*, 2000 by the National Council of Teachers of Mathematics. All rights reserved.

Key Question

What rule was used to sort the shapes, and where do the remaining shapes belong, based on that rule?

Learning Goals

Students will:

- identify the rule used to sort a set of shapes, and
- place additional shapes where they belong based on that rule.

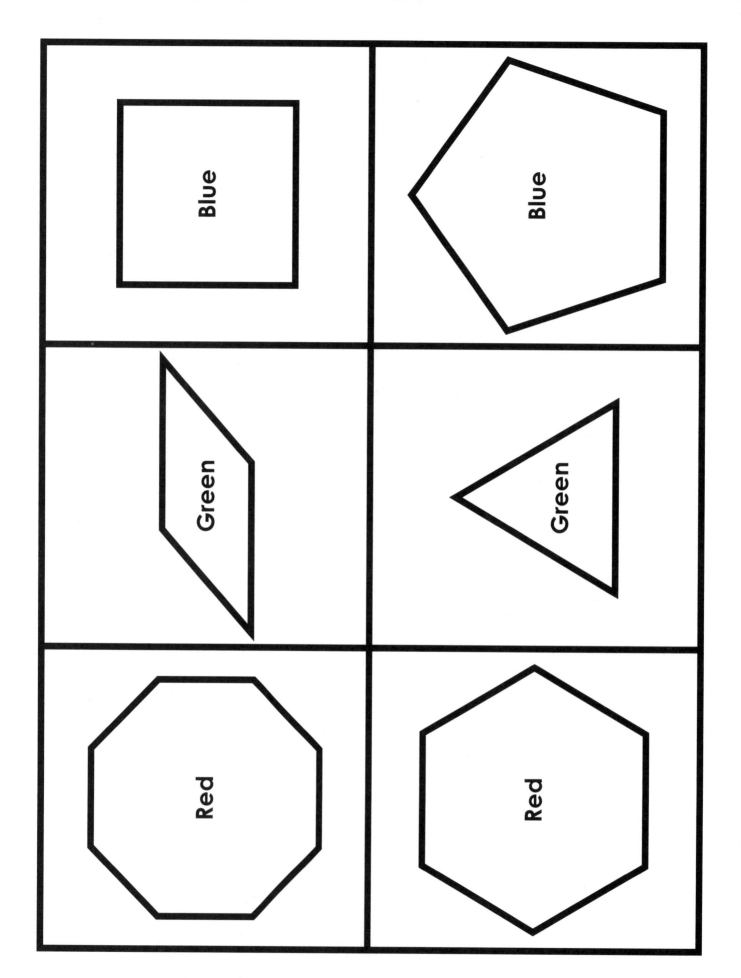

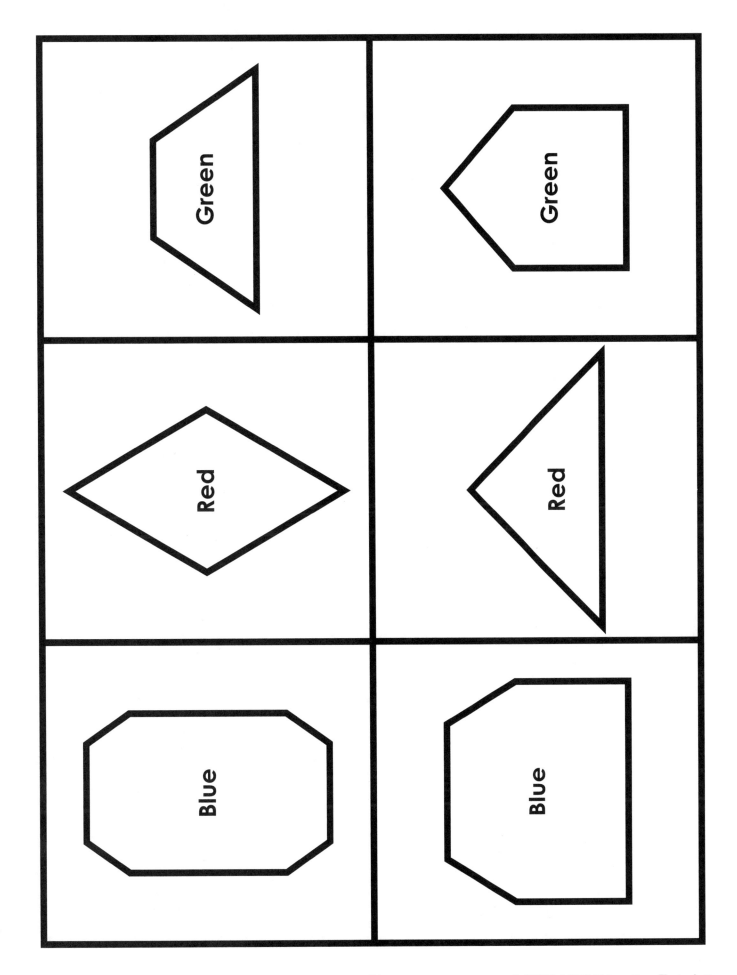

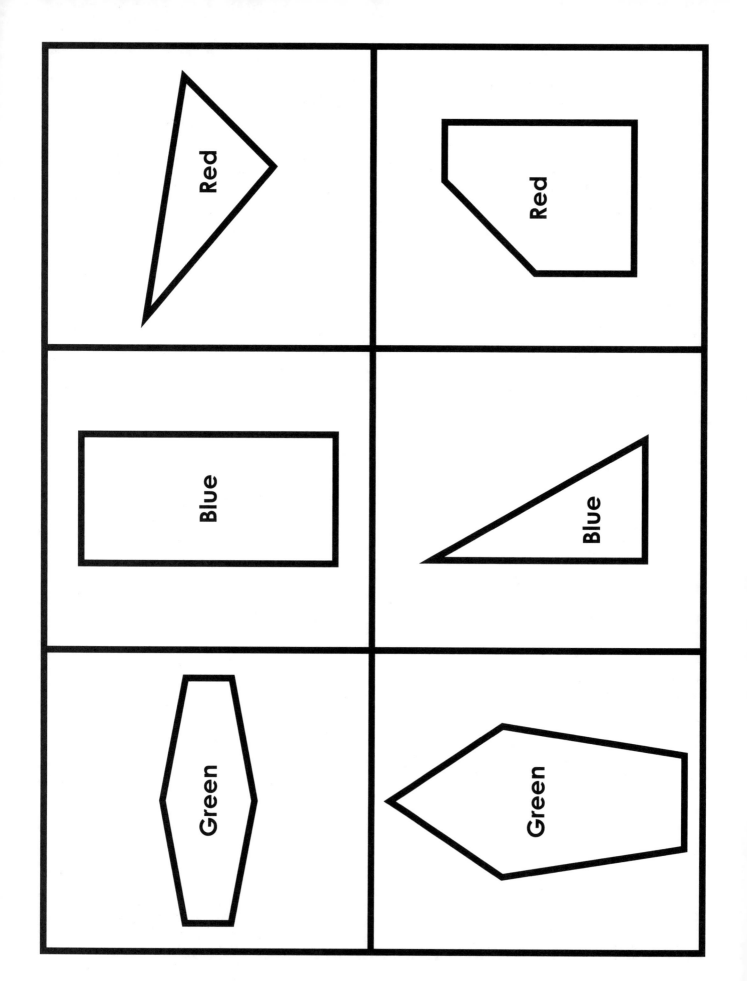

Connecting Learning

1. How were you able to determine the rule used for sorting the shapes?

2. Which sorts were the easiest to identify? …the most difficult?

3. What other ways can you think of to sort the shapes?

4. What does it mean when there are two intersecting circles on a Venn diagram?

Purpose of the Game
Students will identify pairs of geometric shapes based on a variety of attributes.

Materials
Cards from *Sorting Shapes*
Paper to keep score

Management
1. Each group of three to five students needs one set of cards from the activity *Sorting Shapes*. They should have completed that activity before doing this one.
2. Students can use scratch paper to keep score. One person in each group will be the scorekeeper.

Rules
1. Shuffle the stack of cards. Lay six cards face up in a three-by-two array.
2. The player with the birthday closest to the day's date begins. Play moves around the circle in a clockwise direction.
3. The first player identifies two cards out of the six that form a pair and states the common characteristic that makes them a pair. Characteristics can include color, shape, kind of angle, odd or even number of sides, etc.
4. If the player is able to make a pair, that player receives one point. If he or she is not able to make a pair, no points are received. (Not being able to find a pair in one round does not prevent a player from making a pair in the following round, should something become obvious.)
5. Play continues until no more pairs can be found by any of the players. Note: The same cards can be used in multiple pairings as long as a different characteristic is used to explain why they are a pair. Also, group members have the right to veto pairings that are either incorrect or not based on the colors or geometric properties of the shapes. (E.g., "The square and the triangle form a pair because they are my two favorite shapes.")
6. This process is repeated with the remaining two sets of six cards. At the end of all three rounds, the player with the most points is the winner.

Shape SHIFTERS

Topic
2-D shapes

Key Question
How can we use a Chinese jump rope to model different shapes?

Learning Goals
Students will:
- construct two-dimensional geometric shapes using Chinese jump ropes, and
- compare and contrast two-dimensional geometric shapes.

Guiding Document
NCTM Standards 2000*
- *Describe attributes and parts of two- and three-dimensional shapes*
- *Recognize, name, build, draw, compare, and sort two- and three-dimensional shapes*
- *Create mental images of geometric shapes using spatial memory and spatial visualization*
- *Recognize and represent shapes from different perspectives*
- *Organize and consolidate their mathematical thinking through communication*

Math
Geometry
   2-D shapes
Spatial sense

Integrated Processes
Observing
Comparing and contrasting
Communicating
Relating

Materials
For the class:
   The Greedy Triangle
      (see *Curriculum Correlation*)
   chart paper

For each group:
   Chinese jump rope (see *Management 3*)

Background Information
Geometry for young children begins with observing, classifying, describing, and naming shapes. Primary students begin by using their own vocabulary to describe objects, often telling how they are alike or different. Teachers should help their students to gradually incorporate conventional terminology into their descriptions of these shapes.

This activity is designed to let children explore shape through literature, body movements, and manipulatives, while providing an opportunity to develop the conventional terminology needed to describe the attributes (number of sides, number of vertices, intersecting lines, parallel lines) of two-dimensional objects.

Management
1. This activity will take several days to complete.
2. Divide students into groups of four when using the jump ropes.
3. Chinese jump ropes can be purchased at local department stores. If you are unable to get Chinese jump ropes, a six-foot length of elastic will work. Join the ends of the elastic to form a loop.
4. When forming the shapes, it is important for the students to be on the outside of the jump rope. As they pull the jump rope to form the shape, it is easier for them to see the corners if they are not inside the shape.

Procedure
Part One
1. Read *The Greedy Triangle*. Discuss shape words as they appear in the story, focusing on the number of sides and corners for each shape.
2. After reading *The Greedy Triangle*, ask the students to recall some of the shapes from the story. Record their responses on chart paper. Look at each shape word individually and ask the students to tell you everything they remember about that shape, including real-world examples of the shape. (e.g., triangle: three sides, three corners (vertices), pizza is cut into triangles.) Record their responses on a class chart. When looking at the square and rectangle, emphasize that the corners (vertices) are square corners; they are right angles. Show the students the square-edge of a piece of paper. Tell the students that not all shapes have square corners (right angles). Show them how they can use the square corner of a piece of paper as a right angle checker.
3. Discuss ways that the shapes are alike and different.

SHAPES, SOLIDS, AND MORE   © 2009 AIMS Education Foundation

Part Two
1. Gather your class into an open area and have them form a circle by holding hands. Draw attention to the fact that they have just formed a geometric shape that has no straight sides and no corners (vertices).
2. Ask your students to recall the story of *The Greedy Triangle*. Question them about the number of corners (vertices) and sides (edges) that a triangle has. Ask the students how they might make a triangle using their bodies. Allow them to demonstrate the triangles. (Some students will hold their arms above their head and join their hands together to form a point, while others may put their pointer fingers and thumbs together to form a three-sided, three-cornered shape.) Discuss how their triangles are alike and different. [Triangles always have three corners (vertices) and three sides (edges). Sometimes they have different sized corners (vertices) and side (edge) lengths.]
3. Ask the students to describe a triangle. Invite three students to come to the center of the circle. Give them one Chinese jump rope and ask them to make a triangle. Invite the students to make their triangle larger. …smaller.

4. Discuss whether changing the size affected the fact that it was a triangle. [not at all] Ask the students what every triangle must have. [three sides (edges), three corners (vertices)]
5. Divide the class into groups of four and give each group a jump rope.
6. Instruct each group of students to form a triangle. Ask the class what shape they think they would get if they, like the shapeshifter, added one side (edge) and one corner (vertex) to their triangle. [There are several answers that would be correct: a square, a rectangle, a parallelogram, a rhombus, or a trapezoid. They could even make a general quadrilateral (one that has no special attributes other than four sides (edges) and four corners (vertices)).]
7. Discuss the many possible answers.
8. Invite the students to add one side (edge) and one corner (vertex) to their triangle. Instruct them to look at the shapes formed by the other groups. Have the students describe what they observe. [four sides (edges), four corners (vertices)] Ask all groups to make their shapes into squares if they made four-sided shapes other than squares. Have them now describe what they observe. [four square corners (vertices), four sides (edges) of equal length, opposite sides (edges) are parallel, lines intersect at the corners (vertices)]
9. Have them continue constructing shapes by adding one corner (vertex) and one side (edge) each time until they have constructed an octagon. Each time, ask them what they think the new shape will be before they physically add the side (edge) and corner (vertex). Then, have each group make the shapes with the jump rope. After constructing each shape, discuss the number of sides (edges), number of corners (vertices), and the name given to that particular shape.

Connecting Learning
1. Do you think it was easier to make shapes with fewer sides (edges) or more sides (edges)? Why?
2. What were some of the shapes that you were able to make?
3. How were the four-sided shapes the same? …different?
4. What did you discover about triangles? [Triangles can have the same name and yet different sized corners (vertices) and side (edge) lengths.]
5. What did you discover about shapes by constructing them?

Curriculum Correlation
Burns, Marilyn. *The Greedy Triangle*. Scholastic, Inc. New York. 1994.
Unhappy with its shape, a triangle keeps asking the local shapeshifter to add more lines and angles to change its shape thereby introducing various polygons to the reader.

* Reprinted with permission from *Principles and Standards for School Mathematics*, 2000 by the National Council of Teachers of Mathematics. All rights reserved.

Shape SHIFTERS

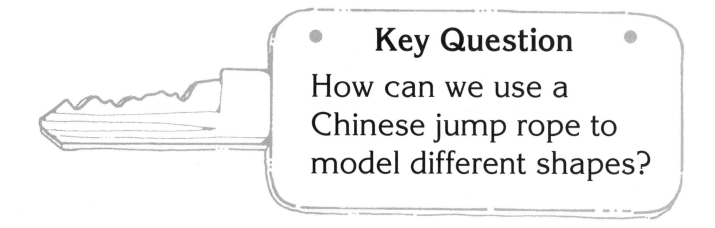

Key Question

How can we use a Chinese jump rope to model different shapes?

Learning Goals

Students will:

- construct two-dimensional geometric shapes using Chinese jump ropes, and
- compare and contrast two-dimensional geometric shapes.

Shape SHIFTERS

Connecting Learning

1. Do you think it was easier to make shapes with fewer sides (edges) or more sides (edges)? Why?

2. What were some of the shapes that you were able to make?

3. How were the four-sided shapes the same? ...different?

4. What did you discover about triangles?

5. What did you discover about shapes by constructing them?

Jump Rope Geometry

Topic
2-D shapes

Key Question
How can making a shape with a jump rope help you understand more about shapes?

Learning Goals
Students will:
- build, compare, and contrast two-dimensional geometric shapes; and
- explore different orientations, sizes, and types of shapes to discover that each two-dimensional shape has certain distinguishable attributes.

Guiding Document
NCTM Standards 2000*
- *Describe attributes and parts of two- and three-dimensional shapes*
- *Investigate and predict the results of putting together and taking apart two- and three-dimensional shapes*
- *Recognize, name, build, draw, compare, and sort two- and three-dimensional shapes*
- *Create mental images of geometric shapes using spatial memory and spatial visualization*
- *Recognize and represent shapes from different perspectives*
- *Organize and consolidate their mathematical thinking through communication*

Math
Geometry
   2-D shapes
Spatial sense

Integrated Processes
Observing
Comparing and contrasting
Communicating
Relating

Materials
For the class:
   Shape Cards (see *Management 3*)

For each group:
   Chinese jump rope (see *Management 4*)

Background Information
One of the underlying concepts found in the geometry standards at the primary level is *spatial sense*. Spatial sense can be described as having an intuitive understanding about two- and three-dimensional shapes and their attributes. While there are several areas of spatial ability that need to be taught and nurtured at the primary level, in this activity we will focus on *spatial visualization* and *spatial memory*. Spatial visualization is the ability to mentally manipulate, rotate, and twist objects. Spatial memory is the ability to create a mental image of an object. It is important to develop this spatial sense because spatial abilities affect other mathematical areas such as problem solving, measurement, and number. Spatial sense can be developed in young learners by allowing for multiple experiences first with concrete and then mental representations of shapes.

Jump Rope Geometry is designed to give the students an opportunity to create mental images of geometric shapes using spatial memory. The students will use the Chinese jump ropes to construct these shapes. In this activity, the students will hold a Chinese jump rope so their hands serve as the vertices of specific shapes. They can then experiment with changing the size and orientations of the shapes.

Management
1. This activity should follow multiple experiences involving the manipulation of concrete two-dimensional shapes.
2. Your class will need to be divided into groups of four when using the jump ropes.
3. Duplicate and laminate the shape cards.
4. Chinese jump ropes can be purchased at local department stores. If you are unable to get Chinese jump ropes, a six-foot length of elastic will work. Join the ends of the elastic to form a loop.
5. When forming the shapes, it is important for the students to be on the outside of the jump rope. As they pull the jump rope to form the shape, it is easier for them to see the corners if they are not inside the shape.

Procedure
1. Review the geometric shapes with your students, allowing them to identify the names and attributes of the square, rectangle, rhombus, triangle, and parallelogram.
2. Gather your class into an open area. Give each group of four a Chinese jump rope.

SHAPES, SOLIDS, AND MORE

3. Hold up the card with the triangle on it. Ask the groups of students to use the jump rope to construct the shape illustrated on the card. After the shapes are made, discuss ways in which the triangles are alike and different. [The triangles could have different-sized corners (vertices) and side (edge) lengths, but they must always have intersecting lines, three sides (edges), and three corners (vertices).]

4. Have the students change their triangles so that the sides (edges) are all equal in length. Ask them to make a triangle that has two sides (edges) that are equal, and then a triangle that has no sides (edges) equal. Discuss what they did to change their triangles and how they looked different.

5. Instruct the students to turn their triangles in different directions so that they can see them from many orientations.

6. Hold up the card with the square on it. Direct the students to change their triangles into squares. Ask them to explain what they had to do to the triangle to change it. Direct the students to make a smaller square, then a larger square. Have the class observe the shape and describe its attributes each time.

7. Question the students about the attributes of a rectangle. [opposite sides are parallel and equal in length, four square corners (vertices)] Display the rectangle card. Ask the students to construct the rectangle. Have them observe the shape from several orientations.

8. Continue displaying the shape cards, discussing the attributes of each shape and instructing the students to construct the shape until all cards have been completed.

Connecting Learning

1. How does changing a shape's orientation affect the shape? [It doesn't.]
2. Name several four-sided shapes. [trapezoid, parallelogram, square, rectangle]
3. How do you know that you formed the shape correctly? [I counted the number of sides (edges) and checked to see if the sides (edges) were intersecting, etc.]
4. Were some of the shapes easier to make than others? Explain.
5. How are the parallelogram, square, rhombus, rectangle, and trapezoid the same? What makes each shape different than all of the others?
6. How did constructing the shapes with the jump ropes help to give you a better understanding of the shapes?

Extension

Create a shape museum outside by using tent stakes to stake several jump rope shapes to the ground.

* Reprinted with permission from *Principles and Standards for School Mathematics,* 2000 by the National Council of Teachers of Mathematics. All rights reserved.

Jump Rope Geometry

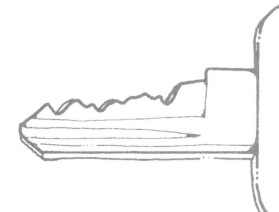

Key Question

How can making a shape with a jump rope help you understand more about shapes?

Learning Goals

Students will:

- build, compare, and contrast two-dimensional geometric shapes; and
- explore different orientations, sizes, and types of shapes to discover that each two-dimensional shape has certain distinguishable attributes.

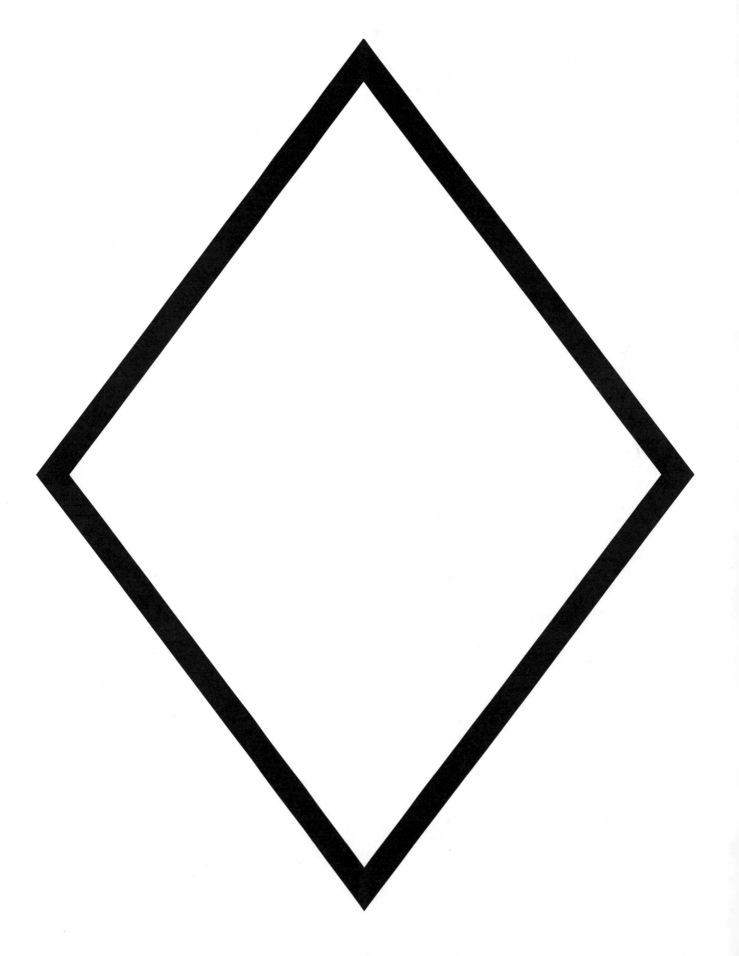

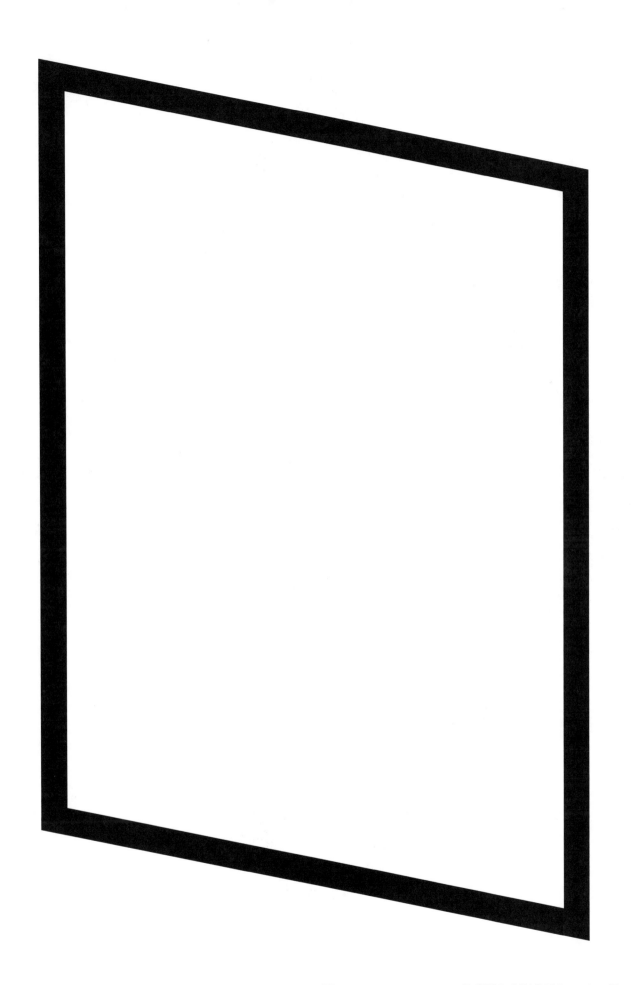

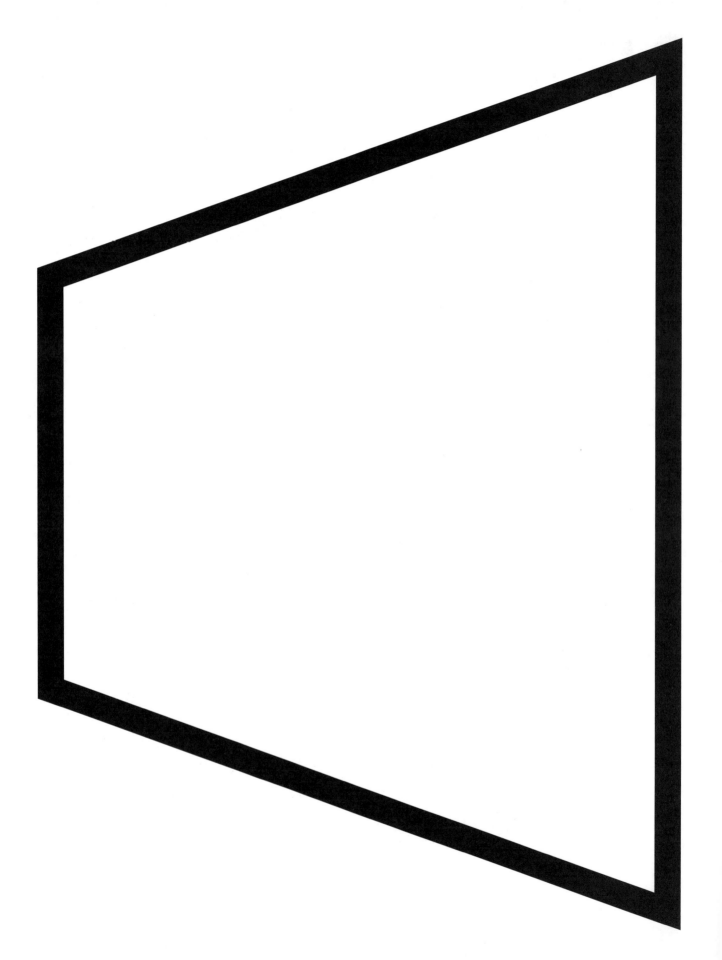

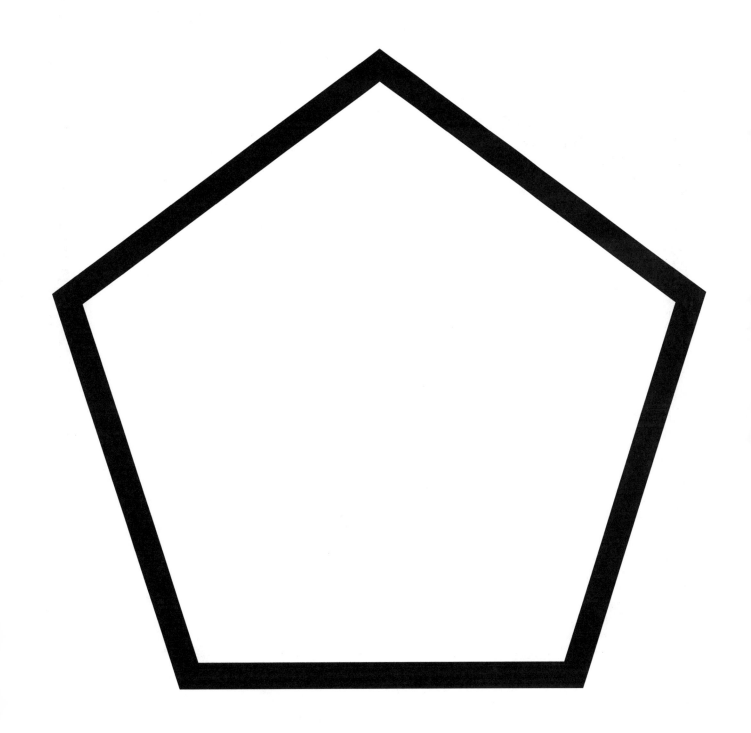

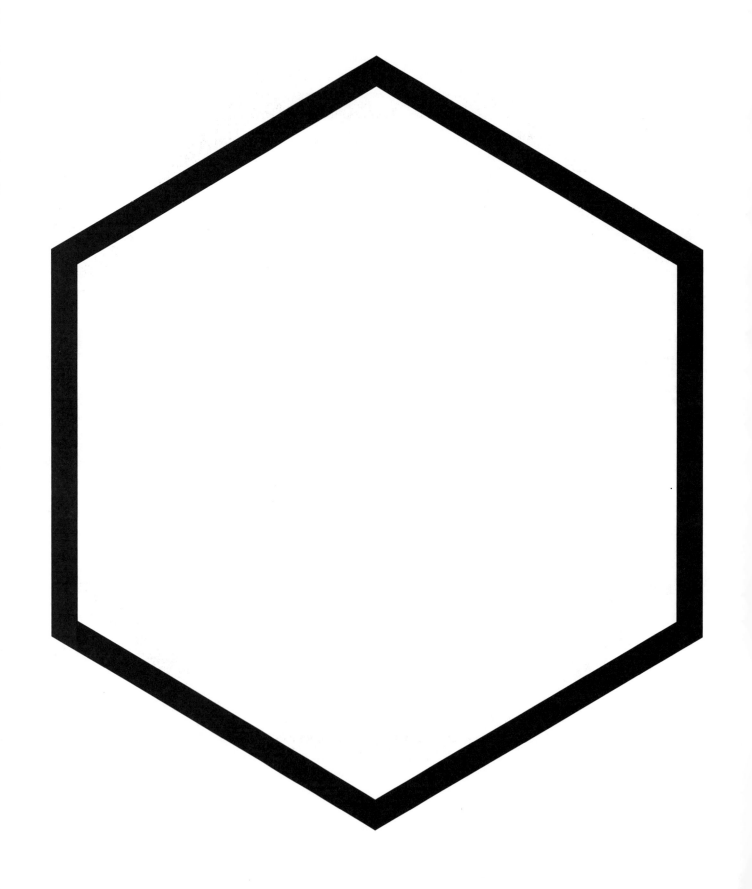

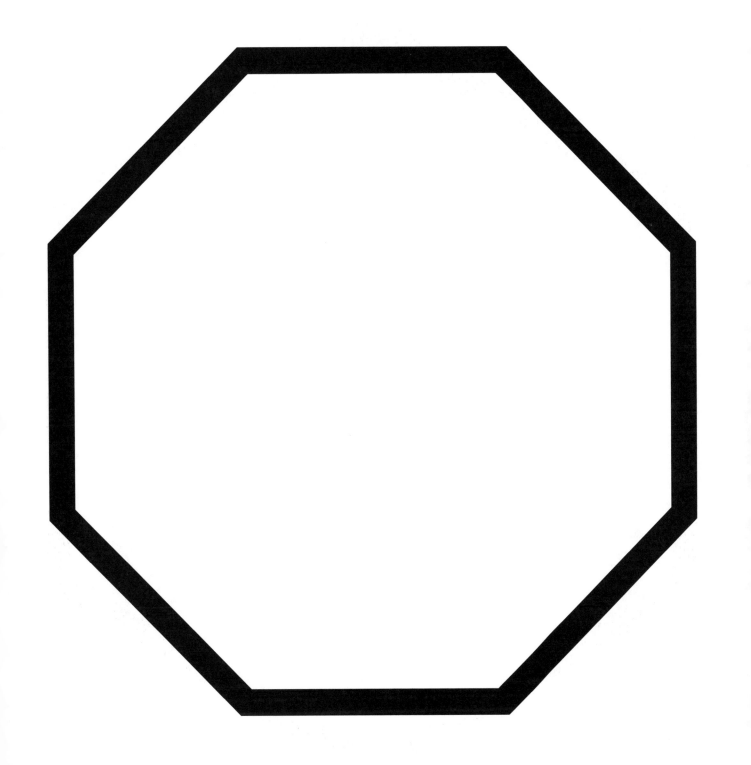

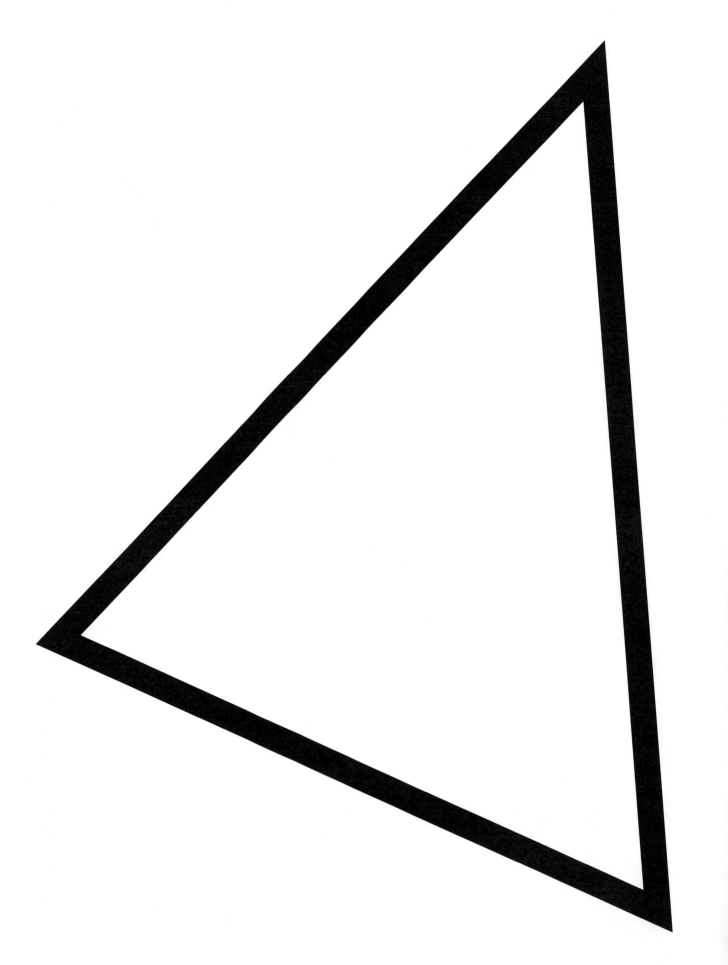

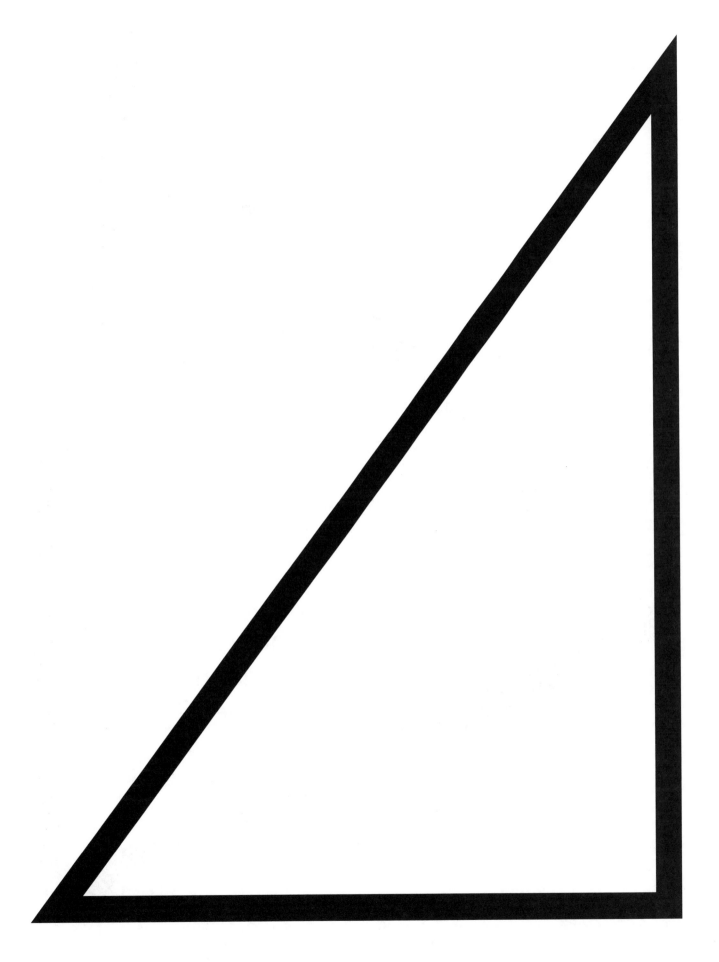

Jump Rope Geometry

Connecting Learning

1. How does changing a shape's orientation affect the shape?

2. Name several four-sided shapes.

3. How do you know that you formed the shape correctly?

4. Were some of the shapes easier to make than others? Explain.

5. How are the parallelogram, square, rhombus, rectangle, and trapezoid the same? What makes each shape different than all of the others?

6. How did constructing the shapes with the jump ropes help to give you a better understanding of the shapes?

Topic
Polygons

Key Question
What are polygons?

Learning Goals
Students will:
- define the term *polygon*, and
- sort shapes into polygons and non-polygons.

Guiding Documents
Project 2061 Benchmarks
- *Numbers and shapes—and operations on them—help to describe and predict things about the world around us.*
- *Many objects can be described in terms of simple plane figures and solids. Shapes can be compared in terms of concepts such as parallel and perpendicular, congruence and similarity, and symmetry. Symmetry can be found by reflection, turns, or slides.*

NCTM Standards 2000*
- *Identify, compare, and analyze attributes of two- and three-dimensional shapes and develop vocabulary to describe the attributes*
- *Build and draw geometric objects*

Math
Geometry
    2-D shapes
        polygons

Integrated Processes
Observing
Comparing and contrasting
Relating

Materials
Shape cards
Shape labels

Background Information
The study of two-dimensional shapes generally begins with students first classifying shapes as polygons (many angles)/non-polygons (no angles). As their geometry knowledge develops, they will progress to classifying polygons as triangles, quadrilaterals, pentagons, hexagons, or octagons based on their number of sides. Since many of these terms are new for students, practice and review will be needed for mastery of these terms.

Management
1. Prior to teaching this lesson, copy the *Polygon, Non-polygon* labels and shape pictures onto card stock and laminate for extended use.

Procedure
1. Tell the class that they are going to learn some new geometry terms. Write the word *polygon* on the board. Explain that *poly* means "many" and *gon* means "angles." Ask the students to define *polygon*. [many angled figure.]
2. Display pictures of a circle, triangle, square, and rectangle on the board. Ask the students to help you identify which of the displayed shapes are polygons. Place the *Polygon* label on the board and position the triangle under it. Discuss why the shape is a polygon.
3. Question the students about why the circle would not be considered a polygon. [It has no angles.] Explain that objects that are not polygons are called non-polygons (not polygons). Place the *Non-polygon* label on the board and move the picture of the circle under it.
4. Show students the additional shape cards and invite them to place them under the label that best describes them. Each time a shape is placed, ask the students to tell why it belongs where they placed it.
5. End with a review of the terms *polygon* and *non-polygon*.

Connecting Learning
1. What does *poly* mean? [many] What is the meaning of *gon*? [angles]
2. Do all polygons have straight sides? [Yes.] How do you know this? [A polygon is a shape with many angles; to have angles, you must have straight sides.]
3. Give an example of a shape that is not a polygon. [circle] Why is it not a polygon? [no angles]
4. What other ways could we have sorted these shapes?

* Reprinted with permission from *Principles and Standards for School Mathematics,* 2000 by the National Council of Teachers of Mathematics. All rights reserved.

Polygon or Non?

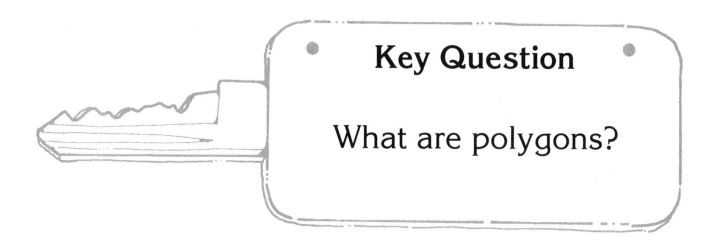

Key Question

What are polygons?

Learning Goals

Students will:

- define the term *polygon*, and
- sort shapes into polygons and non-polygons.

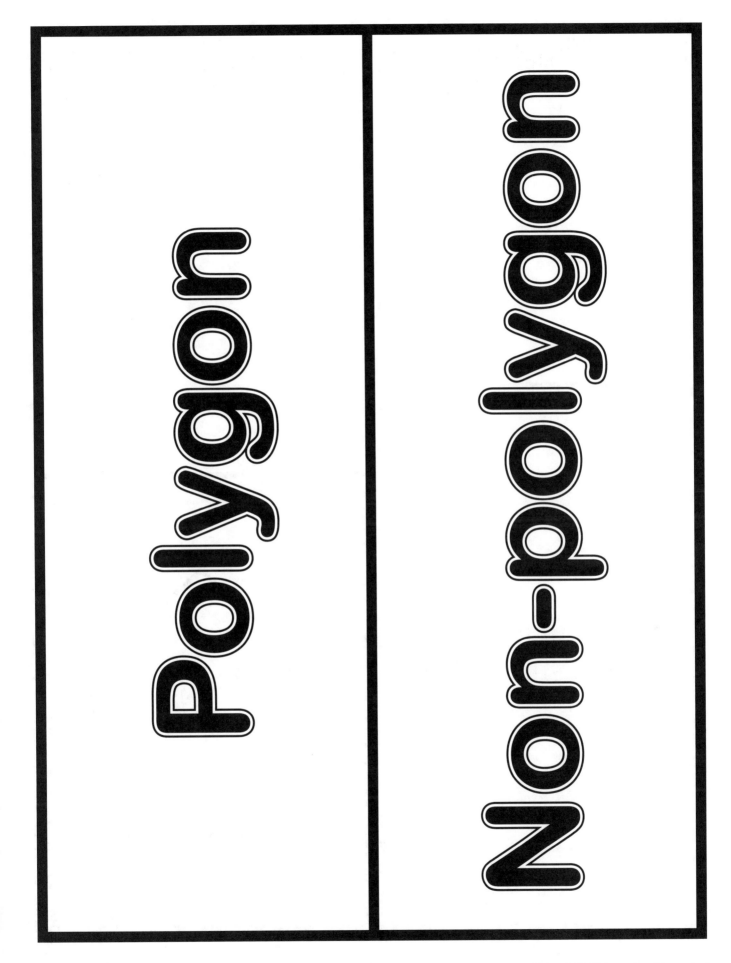

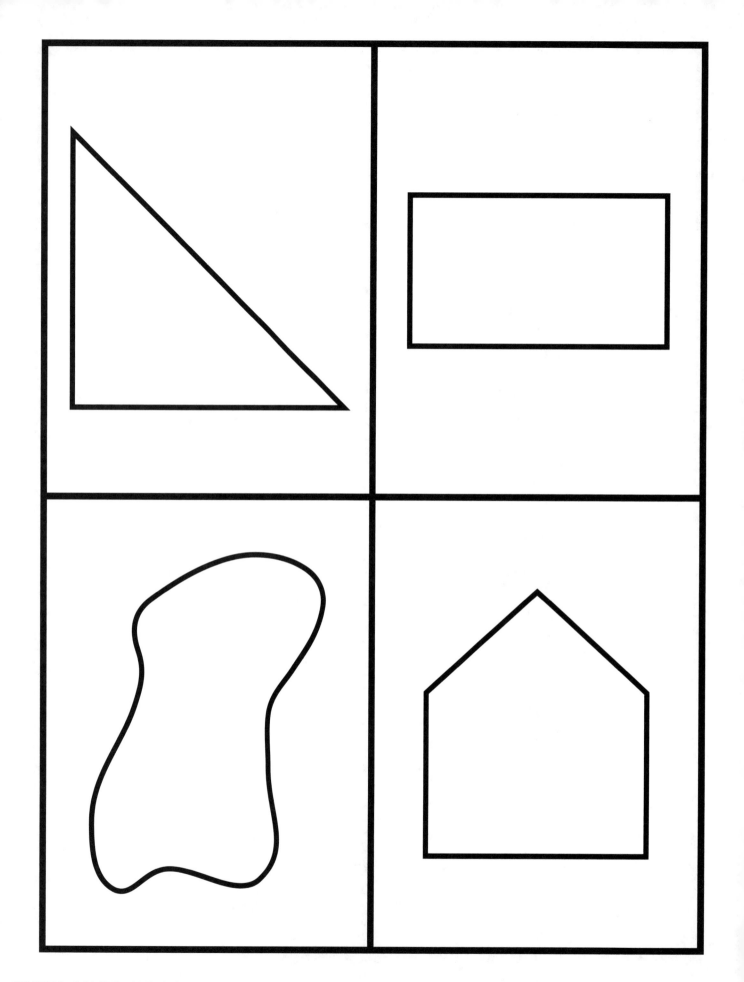

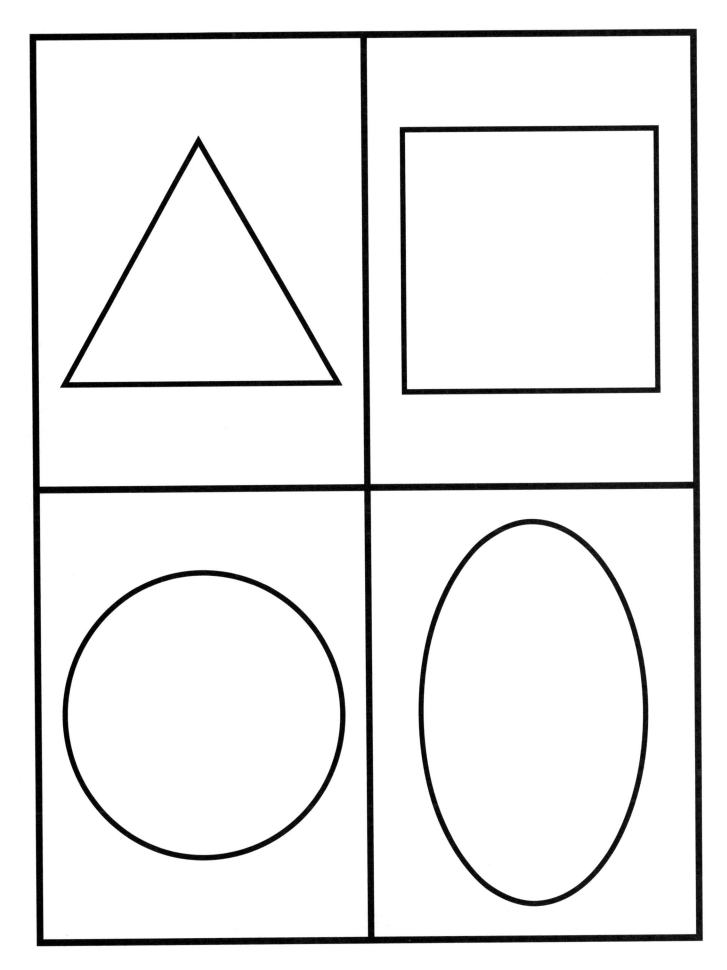

SHAPES, SOLIDS, AND MORE © 2009 AIMS Education Foundation

Polygon or Non?

Connecting Learning

1. What does *poly* mean? What is the meaning of *gon*?

2. Do all polygons have straight sides? How do you know this?

3. Give an example of a shape that is not a polygon. Why is it not a polygon?

4. What other ways could we have sorted these shapes?

Pondering Polygons

Topic
Polygons

Key Questions
1. How do polygons compare?
2. How do we classify polygons?

Learning Goals
Students will:
• explore triangles and quadrilaterals, and
• identify and sort specific polygons.

Guiding Documents
Project 2061 Benchmarks
• *Numbers and shapes—and operations on them—help to describe and predict things about the world around us.*
• *Many objects can be described in terms of simple plane figures and solids. Shapes can be compared in terms of concepts such as parallel and perpendicular, congruence and similarity, and symmetry. Symmetry can be found by reflection, turns, or slides.*

NCTM Standards 2000*
• *Identify, compare, and analyze attributes of two- and three-dimensional shapes and develop vocabulary to describe the attributes*
• *Build and draw geometric objects*

Math
Geometry
    2-D shapes
        polygons

Integrated Processes
Observing
Comparing and contrasting
Relating

Materials
For each student:
    geoboard
    #19 rubber band
    student pages
    Pick a Polygon cards

For the class:
    meter tape
    triangle cards
    quadrilateral cards

Background Information
In the activity *Polygon or Non?* students developed an understanding of polygons and non-polygons. To further build on that geometric knowledge, students will now look at classifications of polygons such as triangles and quadrilaterals. They will soon discover that polygons with the same number of sides may look quite different but are still classified the same way. For example, all four-sided polygons are quadrilaterals regardless of their side lengths. Therefore, squares, rectangles, trapezoids, rhombi, as well as the following shapes, are all quadrilaterals.

Since many of the terms in this activity are new to students, practice and review will be needed for mastery of these terms.

Once the students have an understanding that all three-sided polygons are classified as triangles and four-sided polygons are quadrilaterals, they can then begin classifying polygons by their side lengths and angle measures.

Management
1. Make one set of *Pick a Polygon* cards for each pair of students. Copy two of each page onto card stock and laminate for extended use.
2. Copy a set of triangle and quadrilateral cards on card stock and laminate for extended use.

Procedure
Part One—Triangles
1. Review the terms *polygon* and *non-polygon*. Explain to the class that in this lesson, they will look closely at some of the polygons explored in the previous lesson.

SHAPES, SOLIDS, AND MORE © 2009 AIMS Education Foundation

2. Give each student a geoboard, a #19 rubber band, and the first recording page. Have everyone make a triangle on the geoboard and record it on the student page. Ask students to describe their triangles—three equal sides, one short and two long, etc.
3. Ask the students to compare their triangles to their neighbors' triangles and report to the class any similarities and differences. Bring attention to the fact that all triangles have three sides and three angles but the length of the sides may be different.
4. Discuss how all three-sided figures are specific polygons called triangles. Tell them that the word *triangle* can be broken down into two parts—*tri* which means three—and *angle*. Triangles are polygons with three angles. Have students count the angles on their triangles.
5. Draw a right triangle on the board. Explain to the students that it is a triangle, which is a polygon, and that it is specifically a right triangle. Encourage students to tell why they think it is a right triangle. [It has a right angle.] Have them build a right triangle on their geoboards and record it on the student page.
6. Have students build and draw as many different three-sided polygons as they can in the space provided on the student page.
7. Place the triangle cards on the board. Ask students to suggest how the polygons could be sorted. [length of sides] Invite students to measure the triangles and label the length of each side. Sort the triangles using various rules.

Part Two—Quadrilaterals
1. Review what a polygon is and how a shape can be both a triangle and a polygon.
2. Distribute rubber bands, geoboards, and the second recording page.
3. Ask students to make a four-sided figure on their geoboards. Question the students about whether the four-sided shape they made is a polygon.
4. Explain that all four-sided polygons are called *quadrilaterals*. Tell them that *quad* means four and *laterals* are sides. Quadrilaterals are four-sided figures. Have students share what shape they made. Talk about how a square, rectangle, trapezoid, etc., are all polygons with four sides so they are also called quadrilaterals. Have students build and draw as many quadrilaterals as they can in the space provided for them on the student page. Invite students to share the four-sided shapes that they drew.
5. Place the quadrilateral cards on the board. Ask for suggestions as to how they could be sorted. Sort the quadrilaterals using the various rules.

Part Three—Playful practice
1. Review what a polygon is and have students describe how three- and four-sided polygons can have different names.
2. Tell the students that they will be playing a game called *Pick a Polygon*. Explain that the rules for *Pick a Polygon* are very similar to Go Fish.
3. Have each pair of students cut out their set of cards, mix them up, and deal six cards to each student. The remaining cards should be placed face down on table between the two players.
4. Instruct the players to start the game by looking at the cards in their hands and placing any matching pairs of polygons down on the table. Ask them to explain to their partner why the two polygons should be considered a pair. For example, "I put a square and rectangle down because they are both quadrilaterals."
5. Have one student in each group begin play by asking his or her partner for a particular polygon. For example, "Do you have any trapezoids?" Tell the second student to respond by giving the partner a trapezoid, if he/she has one, or by saying, "Pick a Polygon," which means that they should pick a card from the top of the deck. Play continues until one of the two players is out of cards. The winner is the student with the most matches.
6. End with a discussion about how polygons can be described and sorted.

Connecting Learning
1. What makes something a polygon? [It has to have many angles.] Give an example of a shape that is not a polygon. [circle]
2. What other names describe a square? [polygon, quadrilateral]
3. Is a rectangle a polygon? Explain your answer.
4. Explain how a right triangle can be both a triangle and a polygon.
5. How does what we learned about triangles and quadrilaterals apply to pentagons, hexagons, and octagons?

* Reprinted with permission from *Principles and Standards for School Mathematics*, 2000 by the National Council of Teachers of Mathematics. All rights reserved.

Key Questions

1. How do polygons compare?
2. How do we classify polygons?

Learning Goals

- explore triangles and quadrilaterals, and
- identify and sort specific polygons.

Pondering Polygons

These are all polygons: △ ⬠ ◻ △

Make as many triangles as you can. Record a picture of each below.

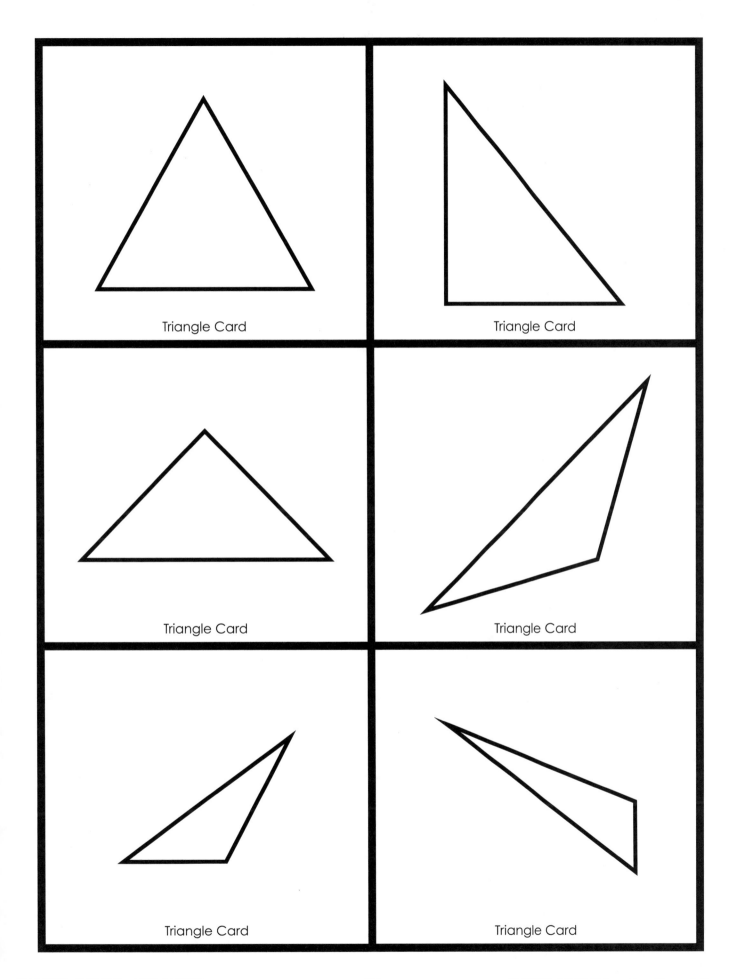

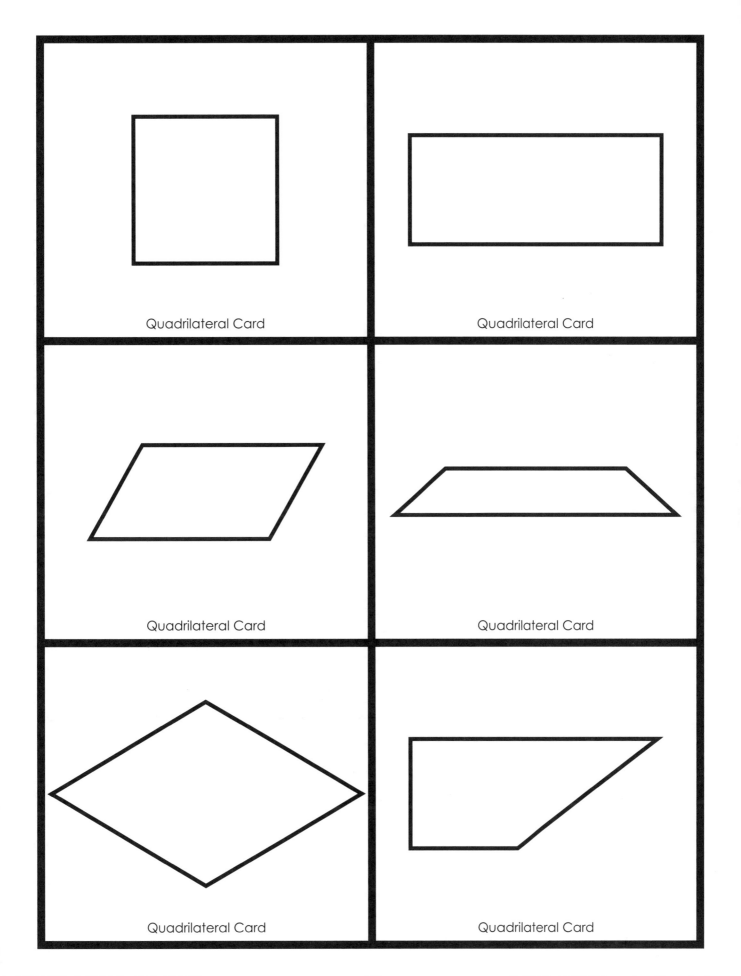

Pick a Polygon | Pick a Polygon
Pick a Polygon | Pick a Polygon
Pick a Polygon | Pick a Polygon
Pick a Polygon | Pick a Polygon

SHAPES, SOLIDS, AND MORE © 2009 AIMS Education Foundation

Pick a Polygon | Pick a Polygon
Pick a Polygon | Pick a Polygon
Pick a Polygon | Pick a Polygon
Pick a Polygon | Pick a Polygon

SHAPES, SOLIDS, AND MORE © 2009 AIMS Education Foundation

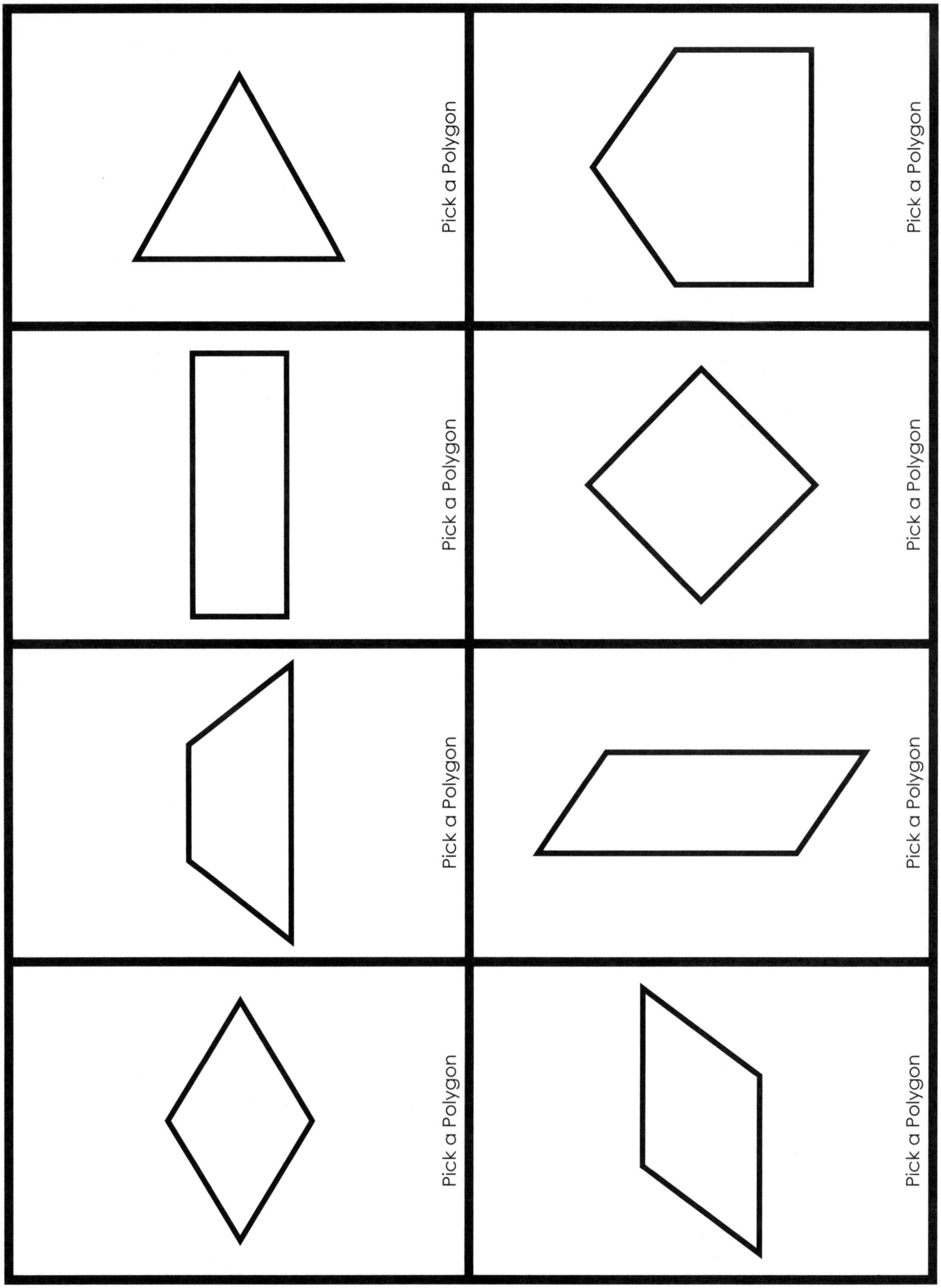

Pick a Polygon

Pick a Polygon

Pick a Polygon

Pick a Polygon

Pick a Polygon

Pick a Polygon

Pick a Polygon

Pick a Polygon

SHAPES, SOLIDS, AND MORE © 2009 AIMS Education Foundation

Pondering Polygons

Connecting Learning

1. What makes something a polygon? Give an example of a shape that is not a polygon.

2. What other names describe a square?

3. Is a rectangle a polygon? Explain your answer.

4. Explain how a right triangle can be both a triangle and a polygon.

5. How does what we learned about triangles and quadrilaterals apply to pentagons, hexagons, and octagons?

Tasty Triangles

Topic
Kinds of triangles

Key Question
What are the different types of triangles?

Learning Goal
Students will use licorice sticks to become familiar with different types of triangles (equilateral, isosceles, scalene).

Guiding Documents
Project 2061 Benchmark
- *Many objects can be described in terms of simple plane figures and solids. Shapes can be compared in terms of concepts such as parallel and perpendicular, congruence and similarity, and symmetry. Symmetry can be found by reflection, turns, or slides.*

NCTM Standard 2000*
- *Classify two- and three-dimensional shapes according to their properties and develop definitions of classes of shapes such as triangles and pyramids*

Math
Geometry
  2-D shapes
    triangles

Integrated Processes
Observing
Comparing and contrasting
Classifying
Collecting and recording data

Materials
Licorice sticks, three per student
Centimeter rulers, one per student
Tasty Triangles journal, one per student

Background Information
The various names given to triangles can be confusing. Learning the language of geometry is similar to learning a foreign language. It takes time and multiple experiences in order for the words to become a part of the working vocabulary. Word confusion is reduced if there is some way to arrange and relate the triangle names.

This activity provides a playful experience in which to discover differences in triangles. Triangles can be named in two ways, by angles and/or by length of sides. This activity focuses on the length of sides for classification. Students will be introduced to the terms equilateral, isosceles, and scalene.

One way to help students remember the order is to have them arrange the words in alphabetical order and then to compare the lengths of sides from largest number of equal sides to fewest.

Equilateral	3 sides equal
Isosceles	2 sides equal
Scalene	0 sides equal

Management
1. Prepare a *Tasty Triangles* journal for each student.
2. Prior to doing the activity, have students wash their hands.
3. Be aware of any students who cannot eat sugar. Drinking straws can be used as a substitute for the licorice sticks Instead of bites, students can cut off bite-sized pieces.

Procedure
1. Ask students what they know about triangles. [They have three sides, three corners, three angles. They are closed.] Explain that students will be using licorice sticks to take a closer look at different kinds of triangles.
2. Give each student three sticks of licorice, a triangle journal, and a ruler. Ask them to make a triangle with the licorice sticks.
3. Have the students measure each of the three sides. Discuss their results. [All sides are equal in length.] Tell the students that when all sides of a triangle are equal in length, it is called an *equilateral triangle*. Make certain that students understand that equilateral comes from the words *equal* and *lateral* for sides. Ask them to draw an equilateral triangle in their journals and to label it. Have them use their own words to write what is special about equilateral triangles.
4. Invite the students to take a medium-sized bite off the end of one of their licorice sticks. Ask them to compare the lengths of their three licorice sticks. [Two are the same length, one is shorter.]
5. Direct them to make a triangle with the three sticks. Have them measure the sides. Discuss their results. Explain that when a triangle has

SHAPES, SOLIDS, AND MORE © 2009 AIMS Education Foundation

two equal sides, it is called an *isosceles triangle*. Ask the students to draw an isosceles triangle in their journal and to label it. Tell them to write the definition of an isosceles triangle in their own words.

6. Have students take two medium-size bites off the end of a whole piece of licorice. Discuss with them how this piece of licorice should be shorter than the other two because one has not been eaten and the second one had just one bite taken off.

7. Have the students use the three pieces to make a triangle. Ask them to measure the lengths of the sides. Discuss their results. Explain that when a triangle has no equal sides, it is a *scalene triangle*. Have students draw a scalene triangle in their journals, label it, and describe what makes it a scalene triangle.

8. Review what an equilateral triangle looks like. Challenge the students to use their licorice sticks to make one. Discuss how they could do this. [Bite off the long sticks until they are the same length as the shortest licorice stick.] Invite them to share how they can prove that the triangle is equilateral and how it compares to the first equilateral triangle they made. [It has three equal sides. The sides of this triangle are shorter than the sides of the first triangle.]

9. Continue challenging the students to make isosceles and scalene triangles by changing the length of their sides.

10. Close with a discussion in which students compare and contrast the three types of triangles they made.

Connecting Learning

1. How did we use the licorice in this activity? [to make triangles]
2. What did we learn about triangles by using the licorice? [Triangles can have sides of different lengths or the same lengths. Triangles are named by the comparison of the lengths of their sides.]
3. How can we tell when the sides of a triangle are equal? [measure them]
4. What are the different types of triangles? [equilateral, isosceles, and scalene]
5. If you have a triangle with all three sides different in length, what type of triangle do you have? [scalene]
6. How can you change an equilateral triangle into an isosceles triangle? [change the length of one side]
7. How can you change an equilateral triangle into a scalene triangle? [make all sides a different length]

* Reprinted with permission from *Principles and Standards for School Mathematics,* 2000 by the National Council of Teachers of Mathematics. All rights reserved.

Tasty Triangles

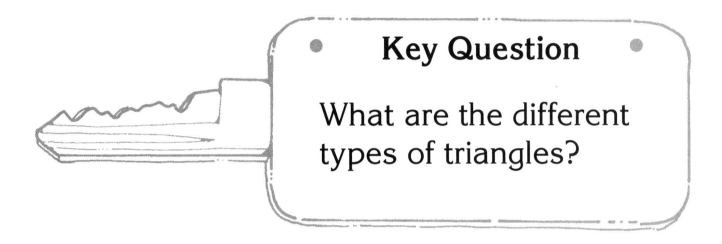

Key Question

What are the different types of triangles?

Learning Goal

use licorice sticks to become familiar with different types of triangles (equilateral, isosceles, scalene).

My first triangle measures _____

It looks like this:

It is called _____

It is special because...

Tasty Triangle Journal

Name _____

My third triangle measures _____

It looks like this:

It is called _____

It is special because...

My second triangle measures _____

It looks like this:

It is called _____

It is special because...

Tasty Triangles

Connecting Learning

1. How did we use the licorice in this activity?

2. What did we learn about triangles by using the licorice?

3. How can we tell when the sides of a triangle are equal?

4. What are the different types of triangles?

5. If you have a triangle with all three sides different in length, what type of triangle do you have?

Tasty Triangles

Connecting Learning

6. How can you change an equilateral triangle into an isosceles triangle?

7. How can you change an equilateral triangle into a scalene triangle?

Picking Out Shapes

Topic
Shape recognition

Key Question
What shapes can be found when we drop a set of Pick Up Sticks®?

Learning Goals
Students will:
- recognize various shapes formed by a random spill of sticks; and
- apply the vocabulary of parallel, intersecting lines, and shape names when identifying the shapes.

Guiding Documents
Project 2061 Benchmark
- *Numbers and shapes can be used to tell about things.*

NCTM Standards 2000*
- *Recognize, name, build, draw, compare, and sort two- and three-dimensional shapes*
- *Recognize and represent shapes from different perspectives*
- *Create mental images of geometric shapes using spatial memory and spatial visualization*
- *Organize and consolidate their mathematical thinking through communication*

Math
Geometry
    lines
    2-D shapes
    spatial sense

Integrated Processes
Observing
Comparing and contrasting
Communicating

Materials
For each group:
    Pick Up Sticks® (see *Management 1*)
    sheet of paper, one per student

Background Information
This lesson is designed to build spatial sense and to increase the use of geometric vocabulary. Spatial sense is increased as the young learners recognize shapes in new settings. Applying the use of the language of geometry helps the students build their understanding of each term. It is assumed that students have learned to name various shapes such as rectangles, squares, triangles, hexagons, octagons, pentagons, etc. The word *quadrilateral* is used in this activity. Quadrilaterals are four-sided polygons. While all squares, rectangles, parallelograms, rhombuses, and trapezoids are regular quadrilaterals, the family of quadrilaterals also includes four-sided shapes with no special attributes. These are called general quadrilaterals. It is necessary to introduce general quadrilaterals here because so many of the shapes that are formed in this activity will fit this category. The students' observation skills are refined as they must look for details, hidden shapes, shapes within shapes, and more.

Management
1. Each group of three students will need 18 Pick Up Sticks®.

Procedure
Part One
1. Introduce the Pick Up Sticks® to the class. Each student should have six sticks.
2. Demonstrate the dropping of the Pick Up Sticks® by holding them upright on the floor in one hand and then letting them fall. Direct the students to do the same.

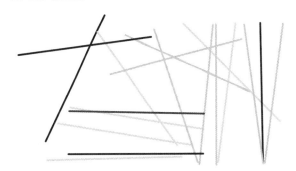

3. Discuss how many sets of parallel lines they see versus the number of intersecting lines. Have the students repeat the stick spill a couple of times, discussing the lines each time. They can use the color of the sticks to aid them in their descriptions.

SHAPES, SOLIDS, AND MORE   © 2009 AIMS Education Foundation

For example: The blue stick is parallel to the red stick. (There will be very few examples of parallel lines.) The yellow stick and the green stick intersect. The yellow stick and the blue stick will intersect at this point.

4. Explain that they are going to play a game called *Picking Out Shapes*. Inform them that they will be playing in small groups of three. Tell them that each person in the group, one at a time, will spill a set of six sticks on the floor in the middle of the group.

5. Ask the students to carefully examine the group stick spill. Challenge them to locate as many different shapes as they can. Have them identify them by calling out the colors of the sticks that form the shapes, or by using another stick to point to the shapes they are naming. For example: I see a red, yellow, and green triangle.

6. Challenge the students to find regular shapes such as squares, rectangles, parallelograms, triangles, etc. The students will find many general shapes that are four-sided but do not have the attributes of a square, rectangle, rhombus, trapezoid, or parallelogram. Introduce or review the word *quadrilateral* by explaining that this is any closed shape with four sides (edges) and four corners (vertices). Explain that even though a square, rectangle, rhombus, trapezoid, and parallelogram are all quadrilaterals, we usually call them by their special names—those they have already learned. Show them a general quadrilateral and tell them that for all the other shapes with four sides (edges) and four corners (vertices), they should call them quadrilaterals. Challenge the students to find pentagons, quadrilaterals, triangles, and more.

7. Optional: To add a number recording aspect to the activity, you may want to assign a point system to certain shapes that are difficult to find when spilling the sticks. For example, a parallelogram may be worth 25 points, a rhombus 15 points, and a triangle only three points. Have the groups keep track of their points on a separate sheet of paper and compare the results at the end of the activity.

8. Discuss what the students have learned about the shapes they found. Ask them to record descriptions and illustrations of these shapes on their papers.

Connecting Learning
1. What are some differences between a square and a general quadrilateral?
2. What are some similarities between a square and a general quadilateral?
3. What shapes did you find the most of?
4. What shapes did you find the least of?
5. Name some of the shapes your group found.
6. What are parallel lines?
7. What are intersecting lines?
8. Which did you see more of, intersecting or parallel lines?

Extension
After spilling a set of sticks, have the students remove one stick from the pile and observe any new shapes that emerge. Have them continue to remove one stick at a time to expose additional shapes.

* Reprinted with permission from *Principles and Standards for School Mathematics,* 2000 by the National Council of Teachers of Mathematics. All rights reserved.

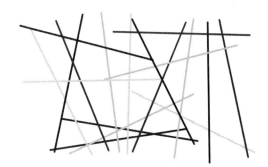

Picking Out Shapes

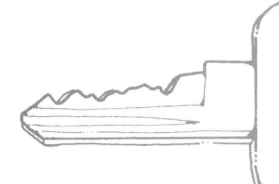

Key Question

What shapes can be found when we drop a set of pick up sticks?

Learning Goals

Students will:

- recognize various shapes formed by a random spill of sticks; and
- apply the vocabulary of parallel, intersecting lines, and shape names when identifying the shapes.

SHAPES, SOLIDS, AND MORE

Picking Out Shapes

Connecting Learning

1. What are some differences between a square and a general quadilateral?

2. What are some similarities between a square and a general quadilateral?

3. What shapes did you find the most of?

4. What shapes did you find the least of?

5. Name some of the shapes your group found.

Picking Out Shapes

Connecting Learning

6. What are parallel lines?

7. What are intersecting lines?

8. Which did you see more of, intersecting or parallel lines?

Creating Congruence

Topic
Congruency

Key Question
What does it mean when two shapes are congruent?

Learning Goals
Students will:
- discover that *congruent* means same size and same shape, and
- make shapes on geoboards congruent to ones made by their partners.

Guiding Document
NCTM Standards 2000*
- *Explore congruence and similarity*
- *Recognize and represent shapes from different perspectives*

Math
Geometry
    2-D shapes
        congruent
        similar

Integrated Processes
Observing
Comparing and contrasting
Applying

Materials
Geoboards
Rubber bands (see *Management 1*)
Congruent cards (see *Management 2*)
Rulers
Student page

Background Information
    Most geometry curriculum asks students to observe characteristics and properties of two-dimensional shapes. Generally in grades K-2, this means for students to identify the number of sides and corners, etc. By third grade, students are also asked to identify whether shapes are *similar* or *congruent*. Congruent objects are exactly the same—they have the same size and the same shape. Similar objects are the same shape, but are not the same size. The mathematical symbol used to identify congruent shapes or line segments is ≅. It is made up of two parts: the ~ which means the same shape and the = means the same size (equal).

    In this activity, students will first observe a series of shapes and generalize what congruent means based on their observations. They will then be asked to make congruent shapes on a geoboard.

Management
1. Ponytail holders can be used instead of rubber bands. They have less elasticity thus making them less tempting to shoot. The downside to using this manipulative is that students will be limited to smaller shapes on the geoboards. Ponytail holders can be purchased at stores that sell items for a dollar or less.
2. Congruent cards can be displayed on the board or copied onto overhead transparencies.

Procedure
1. Display the congruency cards one at a time. After each card is shown, ask the students what they think congruent means. If the students fail to recognize that *congruent* means same size and same shape, you may have to guide them to make that statement.
2. Discuss how the squares on the cards are *similar* because they are the same shape but are not *congruent* because they are not the same size. Invite several students to draw *congruent* shapes on the board. Discuss what makes them congruent. Question students about the possibility of a square and a triangle ever being considered *congruent*. Encourage them to defend their answers.
3. Divide the class into groups of two. Give each pair of students two geoboards, two rubber bands, and two student pages.
4. Have one student in each pair make a shape on his/her geoboard and display it for the other student to see. Give students a minute to observe the size and shape of the figure.
5. Instruct the students who made the shapes to turn their geoboards over. Challenge their partners to make a congruent figure on their own geoboards.
6. When the partners are finished, have the students compare the shapes. Instruct students to record the shapes on the student page and identify whether the two shapes were congruent and explain how they know.
7. Have the students switch roles and repeat the process several times.
8. End with a time for students to share their strategies for making sure the shapes they created were congruent.

SHAPES, SOLIDS, AND MORE

Connecting Learning

1. What does the word *congruent* mean? [same size and same shape]
2. Do all congruent shapes have to be positioned in exactly the same place? Explain. [No. Congruent shapes need only be the same size and shape; orientation does not matter.]
3. Name some classroom objects that are congruent. [Textbooks of the same kind, floor tiles, ceiling tiles, identical bookshelves, identical cupboards, identical windows, etc.]
4. Are all triangles congruent? Why or why not? [No. Triangles come in many sizes and shapes.]
5. Could a square and a triangle ever be considered congruent? Explain your thinking. [No. A square has four sides and a triangle has only three.]
6. Are a penny and a nickel congruent? Why or why not? [No. They are different sizes.]
7. Are all textbooks congruent? Why or why not?

* Reprinted with permission from *Principles and Standards for School Mathematics*, 2000 by the National Council of Teachers of Mathematics. All rights reserved.

Creating Congruence

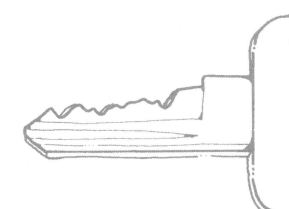

Key Question

What does it mean when two shapes are congruent?

Learning Goals

Students will:

- discover that *congruent* means same size and same shape, and
- make shapes on geoboards congruent to ones made by their partners.

These are not congruent.

What does congruent mean?

These are congruent.

What does congruent mean?

These are not congruent.

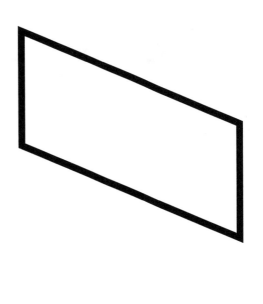
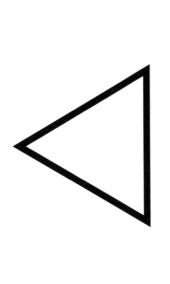

What does *congruent* mean?

These are congruent.

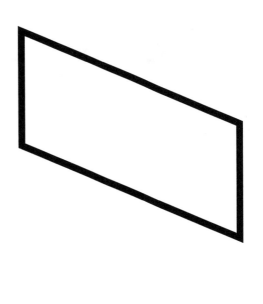

What does *congruent* mean?

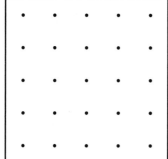 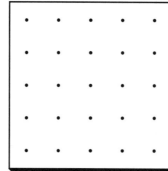 

Were the shapes congruent? Explain. _____

_____

 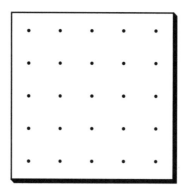 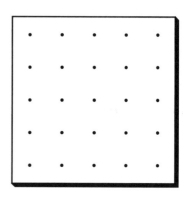

Were the shapes congruent? Explain. _____

_____

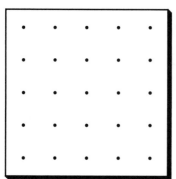 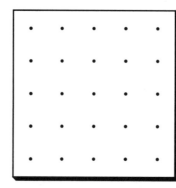 

Were the shapes congruent? Explain. _____

_____

SHAPES, SOLIDS, AND MORE © 2009 AIMS Education Foundation

Creating Congruence

Connecting Learning

1. What does the word *congruent* mean?

2. Do all congruent shapes have to be positioned in exactly the same place? Explain.

3. Name some classroom objects that are congruent.

4. Are all triangles congruent? Why or why not?

5. Could a square and a triangle ever be considered congruent? Explain your thinking.

6. Are a penny and a nickel congruent? Why or why not?

7. Are all textbooks congruent? Why or why not?

Geometric Garden

Topic
2-D shapes

Key Question
What geometric shapes can you identify in your origami scene?

Learning Goals
Students will:
- fold origami tulips, stems, and kites;
- identify the geometric shapes imbedded in their models; and
- describe the characteristics of these geometric shapes.

Guiding Documents
Project 2061 Benchmarks
- Circles, squares, triangles, and other shapes can be found in things in nature and in things that people build.
- Shapes such as circles, squares, and triangles can be used to describe many things that can be seen.

NCTM Standards 2000*
- Recognize, name, build, draw, compare, and sort two- and three-dimensional shapes
- Describe attributes and parts of two- and three-dimensional shapes
- Investigate and predict the results of putting together and taking apart two- and three-dimensional shapes

Math
Geometry
  2-D shapes
  properties of shapes

Integrated Processes
Observing
Comparing and contrasting
Identifying
Recording

Materials
Paper triangles and squares (see *Management 1, 2*)
White construction paper
Transparent tape or glue sticks
Crayons or colored pencils
Student page

Background Information
Origami is a Japanese word that means *to fold paper*. In this activity, students will be able to do simple origami to create a tulip, its stem, and a kite. In so doing, they will encounter numerous 2-D geometric shapes and have an opportunity to identify and describe these shapes. This provides an engaging and meaningful context in which the language of geometry can be explored. Not only will students be able to recognize a triangle as a shape having three sides and three angles, they will be able to hold a triangle in their hands and transform that triangle into another shape. This adds a level of meaning that cannot be achieved with a paper and pencil exercise. Origami has the additional benefits of sharpening fine motor skills and giving students valuable practice at carefully following step-by-step directions. It also provides immediate feedback as students can clearly see at each step whether or not they made the previous fold correctly.

Management
1. Each student will need three pieces of paper in three different colors. The colors should be appropriate for a tulip, its stem, and a kite. The kite will be folded from a square piece of paper, while the tulip and stem will each be folded from a triangular piece of paper. These triangular pieces must be isosceles right triangles, easily made by cutting a square of paper in half along the diagonal. The size of the paper should be determined based on the fine motor skills of your students. The smaller the paper, the harder it is to fold.
2. Students will be best able to see the different shapes they will be folding if the paper is different on the front and back. Most origami paper is "two-sided," meaning that it has a color on one side and white (or a different color) on the other side. Wrapping paper is another easily-obtained paper that is "two-sided." Another option is to use white paper and have students color one side before beginning to fold.
3. You will need to model each fold for students as you go through the instructions. There are many ways to do this, including having several

SHAPES, SOLIDS, AND MORE 115 © 2009 AIMS Education Foundation

adults working with small groups of students, or by using wax paper on the overhead projector to show the entire class at once.
4. This activity is an opportunity for students to practice their geometric vocabulary. The responses indicated in [brackets] are the most advanced and specific responses. You will need to adjust the sophistication and detail of the descriptions based on the abilities of your students. For example, the geometric definition of a *kite* is fairly specific and probably not appropriate for primary students. However, a definition and some questions are included for you as the teacher so that you can guide students' understanding at the appropriate levels.

Procedure
Folding the Tulip
1. Tell students that they are going to make a geometric garden. Distribute a paper triangle to each student for the tulip. Instruct them to place the paper so that the colored side is facing up. Ask them to identify the shape of the paper.
2. Ask students how they know their paper is a triangle. [It has three sides (edges)/corners (vertices).]
3. Have students fold their triangles in half vertically, crease, and unfold. Ask them what shapes their paper has been divided into by the fold. [two triangles]

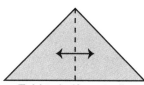 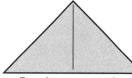
Fold in half vertically   Result: two triangles

4. Ask students to compare these triangles to each other and to the size and shape of the paper. What are the similarities and differences? [The paper has been divided into two congruent triangles. These triangles are similar to the paper—the same shape, but a different size. They have right angles. They are isosceles triangles.]
5. Have students fold the bottom corners of the paper up, starting at the center crease line, as shown below.

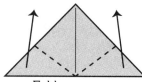
Fold corners up

6. Ask students what shapes they can see now. [two triangles and a kite] Tell them that this is the completed tulip.

Folding the Stem
1. Distribute a second paper triangle to each student for the stem. Instruct them to place the paper so that the colored side is facing down.
2. Have students fold the left side of the triangle down to meet the bottom edge. This will be easier for them if they hold one finger on the corner from which they are folding. Ask them what shapes they can see now. [one big triangle and one small triangle] Discuss how the shapes of the triangles compare to the shape of the paper.

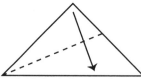 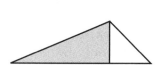
Fold the left side down
to meet the bottom edge

3. Instruct them to repeat this process with the right side of the triangle. Ask them what shapes they can see now.

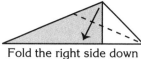 
Fold the right side down
to meet the bottom edge

4. To finish the stem, have students fold along the edge created by the previous fold, as illustrated below. Discuss the shapes created by the folds.

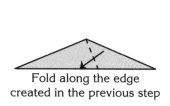 
Fold along the edge
created in the previous step

Folding the Kite
1. Distribute a paper square to each student for the kite. Instruct them to place the paper so that the colored side is facing down. Ask them to identify the shape of the paper.
2. Encourage them to share how they know it is a square. [It has four congruent sides (edges) and four right angles.]

3. Have students fold their squares in half along a diagonal, crease, and unfold.

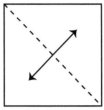
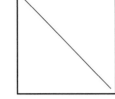

Fold along one diagonal and unfold

4. Ask them to identify the shapes into which the square has been divided. [two isosceles right triangles]
5. Have students fold the top and left sides in to meet at the diagonal crease line. Ask them what shape their paper is now. [a kite]

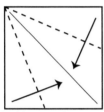
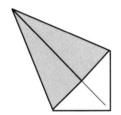

Fold two sides in to meet at the diagonal

6. Invite them to share what shapes they can see in the kite. [triangles] Have them compare the sizes and shapes of the triangles.
7. If desired, allow students to decorate the fronts of their kites.

Creating the Garden
1. Distribute a piece of construction paper to each student. Inform them that they are going to make gardens with the shapes they have folded. Have students tape or glue their tulips and kites to their papers and draw any additional items they may want in their scenes, such as butterflies or clouds. Tell them that they must add at least one new geometric shape to their pictures (such as a round sun or a house with a rectangular door).

Sample garden

2. Give each student a copy of the student page. Have them draw pictures of their scenes in the space provided.
3. Have students identify and label as many different geometric shapes as they can find in their scenes.
4. Close with a final time of discussion.

Connecting Learning
1. How can you tell if a shape is a triangle? [three sides (edges) and corners (vertices)]
2. Do all triangles look the same? [No.] Why or why not? [Any shape with three sides (edges) and three corners (vertices) is a triangle.]
3. Describe some of the different triangles you saw while folding.
4. How can you tell if a shape is a square? [four congruent sides (edges), four right angles]
5. Do all squares look the same? [Yes.] Why or why not? [All squares have four sides (edges) that are the same length and four right angles. The only possible difference between two squares is the length of the sides (edges).]
6. How is a rectangle different from a square? [Two of a rectangle's sides (edges) can be longer than the other two.]
7. Do all rectangles look the same? [No.] Why or why not? [Different rectangles can have different length-to-width ratios.]
8. How can you tell if a shape is a kite? [four sides, opposite angles are congruent (same size and shape), congruent sides are adjacent (next to each other), two sides are shorter than the other two]
9. Do all kites look the same? [No.] Why or why not? [Different kites can have different angles and different lengths of sides.]
10. Describe some of the other geometric shapes you had in your scene.

Extensions
1. Use large pieces of paper and have students fold tulips and kites for display on a bulletin board.
2. Have students fold and/or draw a scene that consists entirely of geometric figures such as triangles, squares, rectangles, circles, etc. Have them label each figure appropriately.
3. Do a geometric shape walk where students go outside and identify geometric shapes in nature and in manufactured objects.

Curriculum Correlation
Dodds, Dayle Ann. *The Shape of Things*. Scholastic, Inc. New York. 1999.
A bright and clever introduction to the concept that shapes make up the world around us. Simple rhymes and bold illustrations help youngsters to see and eventually to draw the world around them.

Murphy, Stuart J. *Circus Shapes: Recognizing Shapes*. HarperCollins. New York. 1998.
Circus animals and performers form basic geometric shapes as they put on a show in this short book with large, simple text and bright pictures.

Pluckrose, Henry. *Shape*. Children's Press. Chicago. 1995.
Photographs of familiar objects introduce basic shapes of squares, circles, rectangles, and triangles with simple, easy-to-read text.

Serfozo, Mary. *There's a Square: A Book About Shapes*. Scholastic, Inc. New York. 1996.
Various shapes such as the square, circle, and triangle invite the readers to search and find examples of their use in this book and in other things seen.

* Reprinted with permission from *Principles and Standards for School Mathematics*, 2000 by the National Council of Teachers of Mathematics. All rights reserved.

Geometric Garden

Key Question

What geometric shapes can you identify in your origami scene?

Learning Goals

Students will:

- fold origami tulips, stems, and kites;
- identify the geometric shapes imbedded in their models; and
- describe the characteristics of these geometric shapes.

Geometric Garden

Draw your garden. Write all the shapes you see.

Connecting Learning

1. How can you tell if a shape is a triangle?

2. Do all triangles look the same? Why or why not?

3. Describe some of the different triangles you saw while folding.

4. How can you tell if a shape is a square?

5. Do all squares look the same? Why or why not?

6. How is a rectangle different from a square?

Connecting Learning

7. Do all rectangles look the same? Why or why not?

8. How can you tell if a shape is a kite?

9. Do all kites look the same? Why or why not?

10. Describe some of the other geometric shapes you had in your scene.

Tangram Tinkerings

Topic
Composition and decomposition of shapes

Key Question
Into what shapes can a square be divided?

Learning Goals
Students will:
- divide a square into a series of smaller shapes,
- identify the smaller shapes into which the square is divided, and
- combine the smaller pieces to make the original square.

Guiding Document
NCTM Standards 2000*
- *Investigate and predict the results of putting together and taking apart two- and three-dimensional shapes*
- *Create mental images of geometric shapes using spatial memory and spatial visualization*

Math
Geometry
  characteristics of 2-D shapes
  shape composition
  shape decomposition
  shape identification
  spatial sense

Integrated Processes
Observing
Comparing and contrasting
Applying

Materials
Square pieces of paper (see *Management 1*)
The Warlord's Puzzle (see *Curriculum Correlation*)
Transparency of tangram directions
Scissors

Background Information
Describing shapes and visualizing what they look like when they are transformed, put together, or taken apart in different ways are important geometry skills for all grade levels. This activity provides an opportunity for students to look at shape decomposition—breaking larger shapes into their smaller component parts, as well as combining smaller shapes to build new shapes. This helps to reinforce properties and characteristics of the shapes. It also facilitates their understanding that the shapes used in this activity can be broken down into smaller components and then those components can be combined to produce other shapes.

Management
1. Each student will need one square piece of paper. The size of the paper should be determined based on the fine motor skills of your students. The smaller the paper, the harder it is to manipulate.
2. Depending on ability level, you may want to lead the class in the construction of the tangram set; otherwise, you may wish to supply the written instructions and have them work on the construction in groups.

Procedures
1. Read *The Warlord's Puzzle* by Virginia Walton Pilegard. Discuss a little of the history of tangrams.
2. Explain to the students that they will be making their own tangram puzzles.
3. Give each student a square piece of paper. Tell the students to fold the square in half along a diagonal. Have them cut along the crease. Ask student to identify the two new shapes. [triangles] Challenge them to combine them to make the original square.
4. Display the tangram directions. Instruct students to take one of the two large triangles, fold it in half as indicated in step two of the instruction sheet, and cut apart along the fold line. Have students identify the three pieces they now have. [one large triangle and two smaller ones] Challenge them to put them together to form the original large square.
5. Have students fold and cut the remaining large triangle as indicated in step three of the instruction sheet. Ask students to identify the four pieces they now have. [three triangles and one trapezoid] Challenge them to put them together to form the original large square.
6. Ask the students to take the trapezoid and fold it in half as indicated in step four of the instruction sheet. When students have cut along the fold line, invite them to identify the five pieces they have. [three triangles and two quadrilaterals] Challenge your students to combine the five pieces to make the original large square.

SHAPES, SOLIDS, AND MORE

7. Instruct students to cut one trapezoid as indicated in step five of the instruction sheet. Have students identify the six pieces they have. [four triangles, one square, one trapezoid] Challenge your students to combine the six pieces to make the original large square.
8. Have students fold the other trapezoid to produce a triangle and a parallelogram as indicated in step six of the instruction sheet. Have students identify the seven pieces they have. [five triangles, one square, one parallelogram] Challenge your students to combine the seven pieces to make the original large square.
9. End the lesson with a discussion about what shapes combine to make other shapes.

Connecting Learning
1. What two shapes can be combined to make a square? [triangles]
2. What shapes can be combined to make a triangle? [two triangles, two triangles and a square, etc.]
3. Into what seven shapes did we break our large square? [five triangles, one square, one parallelogram]
4. How did you decide which shapes to combine to make other shapes?
5. What do you notice about the triangular pieces of your tangram set? [They are similar (same shape, different size), two small triangles make one medium triangle, etc.]

Curriculum Correlation
Pilegard, Virginia Walton. *The Warlord's Puzzle.* Pelican Publishing Co., Inc. Gretna, LA. 2000.

* Reprinted with permission from *Principles and Standards for School Mathematics,* 2000 by the National Council of Teachers of Mathematics. All rights reserved.

Tangram Tinkerings

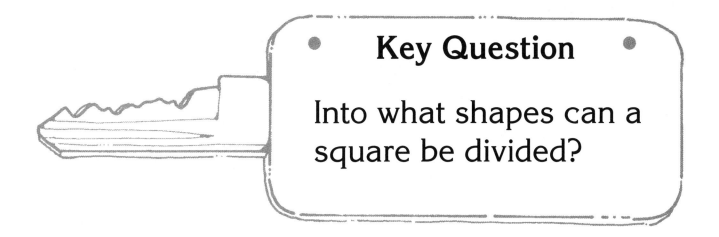

Key Question

Into what shapes can a square be divided?

Learning Goals

Students will:

- divide a square into a series of smaller shapes,
- identify the smaller shapes into which the square is divided, and
- combine the smaller pieces to make the original square.

Tangram Tinkerings

Tangram Cutting Instructions

1. Fold a square of paper in half along a diagonal. Cut along the crease.

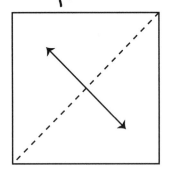 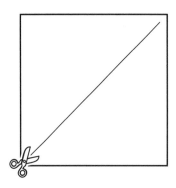 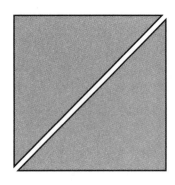

2. Take one of the halves and fold it in half again. Cut along the crease. Set these triangles aside.

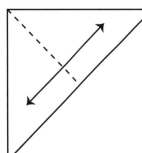 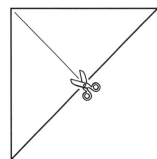 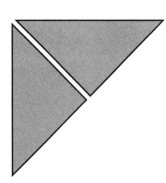

3. Take the remaining large triangle and fold it so that the corner of the right angle touches the midpoint of the opposite side. Cut along the crease to make a small triangle and a trapezoid. Set the triangle aside.

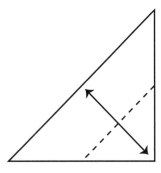 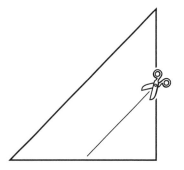 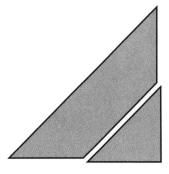

4. Fold the trapezoid in half. Cut along the crease to make two congruent trapezoids.

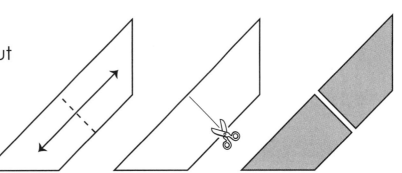

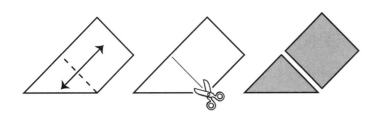

5. Fold and cut one of the trapezoids to make a triangle and a square.

6. Fold and cut the other trapezoid to make a triangle and a parallelogram.

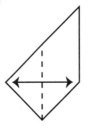 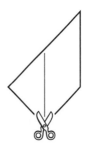 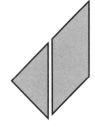

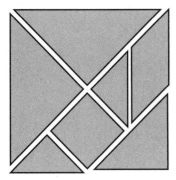

7. These are your seven tangram pieces.

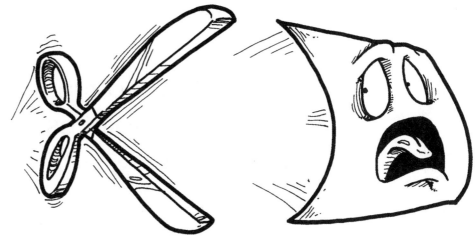

SHAPES, SOLIDS, AND MORE © 2009 AIMS Education Foundation

Tangram Tinkerings

Connecting Learning

1. What two shapes can be combined to make a square?

2. What shapes can be combined to make a triangle?

3. Into what seven shapes did we break our large square?

4. How did you decide which shapes to combine to make other shapes?

5. What do you notice about the triangular pieces of your tangram set?

Cover Alls: Combining Tangrams

Topic
Composing shapes

Key Question
What can we discover about two-dimensional shapes when we combine and cover them?

Learning Goal
Students will investigate the results of putting together two-dimensional shapes.

Guiding Document
*NCTM Standards 2000**
- *Recognize, name, build, draw, compare, and sort two- and three-dimensional shapes*
- *Describe attributes and parts of two- and three-dimensional shapes*
- *Investigate and predict the results of putting together and taking apart two- and three-dimensional shapes*
- *Investigate, describe, and reason about the results of subdividing, combining, and transforming shapes*

Math
Geometry
  2-D shapes
  composing shapes

Integrated Processes
Observing
Classifying
Comparing and contrasting
Communicating

Materials
For each student:
  set of tangram pieces (see *Management 1*)
  journal (see *Management 2*)

Background Information
The focus of a study in geometry for early grades should be on concrete experiences in which young learners observe and describe in their own words the properties of two-dimensional shapes. To develop a more complete understanding of geometric properties, it is important for students to put together (compose) and take apart (decompose) shapes.

In this activity, students will combine shapes in order to create other shapes. Through combining and covering experiences, the students will begin to see relationships among shapes as they explore different orientations of the shapes found in a tangram puzzle. This activity begins slowly with students combining two large triangular pieces to form a parallelogram. It then progresses allowing the children to combine and cover several shapes.

This activity has multiple purposes. It allows students to explore the combining and covering of geometric shapes while providing an opportunity for the teacher to reinforce or assess the use of correct geometric terminology.

Management
1. Each student will need his or her own set of tangram puzzle pieces. Tangram puzzles are available from AIMS (#4180) or students can use the puzzles they created in *Tangram Tinkerings*. You can also copy the page provided and cut out the puzzles for the students. If you will be using paper puzzles, it is recommended that you use card stock and/or laminate the pieces for durability and ease of handling.
2. Make a journal for each student by folding several pieces of blank white paper in half, nesting them together, and stapling along the folded edge.

Procedure
Part One: Composing
1. Distribute a set of tangram shapes and a journal to each student and have students get into pairs.
2. Hold up each of the tangram pieces and review their names. [square, parallelogram, triangle]
3. Review the shapes that have been discussed in previous lessons. [rhombus, circle, rectangle, square, triangle, trapezoid] Tell the students that they will be putting the tangram pieces together to form some of these shapes.
4. Ask the students to find the two large triangles. Demonstrate how the two large triangles can be combined to form a parallelogram. Direct the students to do the same with their pieces. Discuss the side (edge) lengths, corners (vertices), parallel lines, etc. Have the students make a record of this combination by tracing the shapes in their journals.

SHAPES, SOLIDS, AND MORE © 2009 AIMS Education Foundation

5. Have the students work with their partners and use the rest of their tangram pieces to find other possible combinations for using tangram shapes to make a parallelogram. Instruct them to record all combinations discovered and invite groups to share with the class.
6. Tell the students to again find the two large triangles. Ask them what other shapes can be made by combining these pieces. Provide time for exploring, sharing of their discoveries, and recording of the possible combinations before going on.
7. Allow the students to add additional puzzle pieces as you ask them to find shapes that would combine to make a square, trapezoid, rectangle, and triangle. Discuss and record all combinations.

Part Two: Covering
1. Have the students find one large triangle. Demonstrate how two small triangles and one parallelogram can cover the large triangle. Challenge the students to find other combinations that would cover a large triangle.
2. Discuss their findings and have the students record their results in their journals.
3. Urge the students to find combinations that would cover the medium triangle, the square, and the parallelogram. Discuss their findings and have the student record their results in their journals.

Connecting Learning
1. How many combinations did you find that made a square? ...triangle? ...trapezoid? ...etc.?
2. What were some of the shapes that combined to make a square? ...triangle? ...trapezoid? ...etc.?
3. Did any other groups discover combinations that you and your partner did not?
4. Did all the squares you made look the same? ...triangles? ...trapezoids? ...etc.? Explain.
5. What strategies did you use to find combinations that formed new shapes?
6. Is there a way to cover the large triangle with small triangles? Explain.
7. What combinations of shapes could cover a large triangle? ...medium triangle? ...parallelogram? ...square?

Extensions
1. Ask your students to combine the geometric shapes to form pictures.
2. Have your students cover the larger tangram pieces by laying combinations of the smaller pieces on top of them. Remind them that it may be possible to use more than two pieces to cover the shapes.

Curriculum Correlation
Tompert, Ann. *Grandfather Tang's Story: A Tale Told with Tangrams.* Crown Publishers, Inc. New York. 1990.
Grandfather tells a story about shape-changing fox fairies that try to best each other until a hunter brings danger to both of them.

Solutions
Part One
Several possible solutions for composing each shape are shown below.

Parallelogram

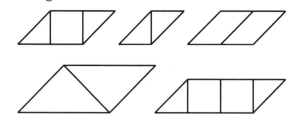

Rectangle

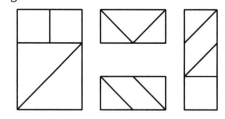

Square

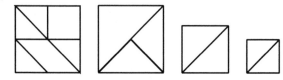

Trapezoid

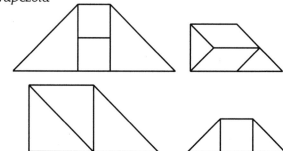

Triangle

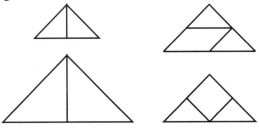

Part Two
The possibilities for covering each of the tangram shapes are shown below.

Large Triangle

Medium Triangle, Square, Parallelogram

* Reprinted with permission from *Principles and Standards for School Mathematics*, 2000 by the National Council of Teachers of Mathematics. All rights reserved.

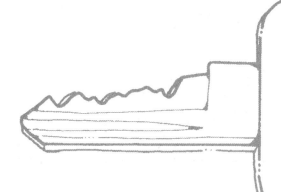

Key Question

What can we discover about two-dimensional shapes when we combine and cover them?

Learning Goal

investigate the results of putting together two-dimensional shapes.

Cover Alls: Combining Tangrams

Connecting Learning

1. How many combinations did you find that made a square? …triangle? …trapezoid? …etc.?

2. What were some of the shapes that combined to make a square? …triangle? …trapezoid? …etc.?

3. Did any other groups discover combinations that you and your partner did not?

4. Did all the squares you made look the same? …triangles? …trapezoids? …etc.? Explain.

Cover Alls: Combining Tangrams

Connecting Learning

5. What strategies did you use to find combinations that formed new shapes?

6. Is there a way to cover the large triangle with small triangles? Explain.

7. What combinations of shapes could cover a large triangle? …medium triangle? …parallelogram? …square?

Animals Take Shape

Topic
Combining shapes

Key Question
How can you combine geometric shapes to make different animals?

Learning Goals
Students will:
- name and recognize the geometric shapes that make up a set of tangrams,
- investigate the results of combining the tangram shapes, and
- use combinations of the tangram shapes to make animals.

Guiding Document
NCTM Standards 2000*
- *Recognize, name, build, draw, compare, and sort two- and three-dimensional shapes*
- *Investigate and predict the results of putting together and taking apart two- and three-dimensional shapes*
- *Build new mathematical knowledge through problem solving*
- *Apply and adapt a variety of appropriate strategies to solve problems*

Math
Geometry
    2-D shapes
        combining shapes
Problem solving

Integrated Processes
Observing
Identifying
Recording

Problem-Solving Strategies
Guess and check
Use manipulatives

Materials
Student pages
Tangram puzzle pieces (see *Management 1*)

Background Information
The tangram puzzle is one that has been around for a very long time, with a multitude of variations. The puzzle consists of five isosceles right triangles (two large, one medium, and two small), a square, and a parallelogram. All of the shapes (except the two large triangles) can be combined with others in some way to make one or more of the larger shapes. For example, the two small triangles can be put together to make the medium-sized triangle, the square, or the parallelogram. The traditional challenge is to use all of the shapes to make a square, but the pieces lend themselves to creating many other figures as well.

This activity gives young learners the opportunity to use their problem-solving skills as they try to put the tangram pieces together to form a variety of animal shapes. The strategy of guessing and checking is very appropriate here, and is the strategy most often used by adults and children alike when confronted with problems of this nature.

There are many different challenges presented, some more difficult than others, so that students at every level can be given appropriate problems. There is also the opportunity to build and develop geometric vocabulary and spatial sense as the students name the shapes they are working with and learn the results of moving them around and putting them in different orientations.

Management
1. Each student will need his or her own set of tangram puzzle pieces. Tangram puzzles are available from AIMS (#4180) or students can use the puzzles they created in *Tangram Tinkerings*. You can also copy the page provided and cut out the puzzles for the students. If you will be using paper puzzles, it is recommended that you use card stock and/or laminate the pieces for durability and ease of handling.
2. There are a total of 12 challenges given, all at different levels of difficulty. All students should begin with the first page and only move on to the remaining pages if appropriate.
3. To record their solutions, have students trace around their tangram pieces inside the frame of the shape.

SHAPES, SOLIDS, AND MORE   © 2009 AIMS Education Foundation

Procedure

1. Give each student a set of tangram puzzle pieces. Have students name and describe each kind of shape in the set. Ask them what they can tell you about each of the shapes.
2. Allow time for some free exploration with the shapes so that students can see how they fit together and the relationship between the sizes of the different pieces.
3. Distribute the first student page and be sure that everyone understands the challenges. Allow time for students to solve each of the three puzzles.
4. Have students share their solutions with the class. Encourage them to make use of geometric vocabulary as they describe how they put the pieces together.
5. As appropriate, distribute the remaining student pages. For this portion, students may be less frustrated if they work in groups to solve the puzzles.
6. Close with a time of class discussion and sharing.

Connecting Learning

1. What shapes are your puzzle pieces? [triangles, square, parallelogram]
2. What do you notice about all of the triangle pieces? [Various. They are all the same shape, just different sizes; two little ones make one medium one; etc.]
3. What else do you notice about the puzzle piece shapes? [Various. Two small triangles make a square, the square and two small triangles make a big triangle, etc.]
4. What shapes could you make by combining the two small triangles? [square, parallelogram, larger triangle]
5. What shapes could you make by combining the three smallest triangles? [larger triangle, parallelogram, square]
6. How did you decide which puzzle pieces to use for each puzzle? Could you ever use different pieces to solve the same puzzle? (Several of the puzzles have multiple solutions.)
7. Which puzzle was the easiest for you to solve? Why?
8. Which puzzle was the most difficult for you to solve? Why?

Extensions

1. Have students create their own tangram puzzles and trade them with other classmates to solve.
2. Have students work in groups and use two or three sets of tangrams to see what shapes can be created.
3. Use the tangram pieces to explore concepts of symmetry, flips, slides, turns, etc.

Solutions

Butterflies

Duck/Penguin

Elephant/Cat

Bat/Fox

Moth Swan

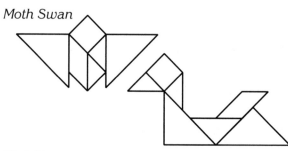

Bird/Goose

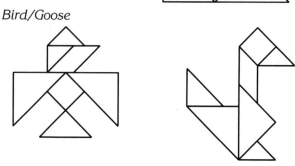

* Reprinted with permission from *Principles and Standards for School Mathematics*, 2000 by the National Council of Teachers of Mathematics. All rights reserved.

Animals Take Shape

Key Question

How can you combine geometric shapes to make different animals?

Learning Goals

Students will:

- name and recognize the geometric shapes that make up a set of tangrams,
- investigate the results of combining the tangram shapes, and
- use combinations of the tangram shapes to make animals.

Tangram Pieces

Copy this page onto card stock and cut out the puzzle pieces. Each student will need one set.

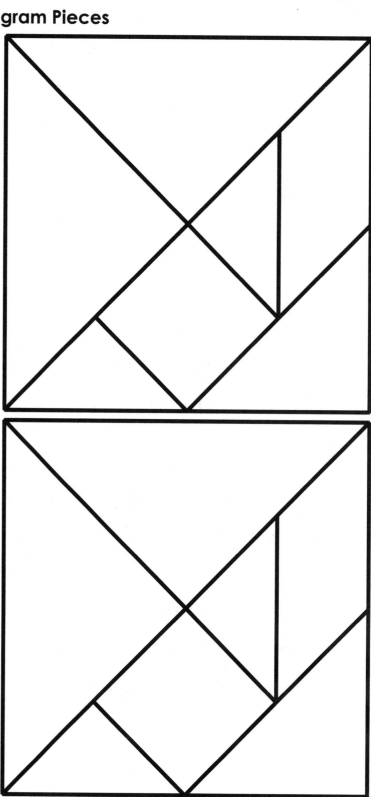

SHAPES, SOLIDS, AND MORE

Animals Take Shape

Make this butterfly using two shapes.

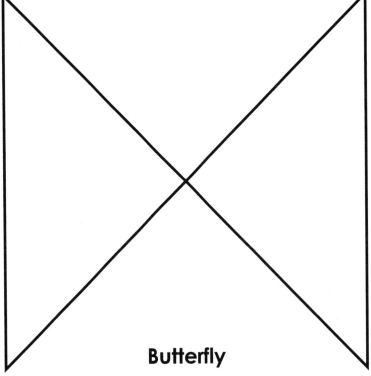

Butterfly

Now use four shapes. Show how you did it.

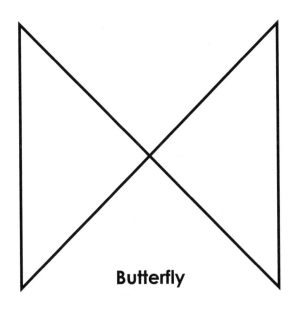

Butterfly

Make the smaller butterfly using three shapes.

Show how you did it.

Animals Take Shape

Make this duck using four shapes.

Duck

Show how you did it.

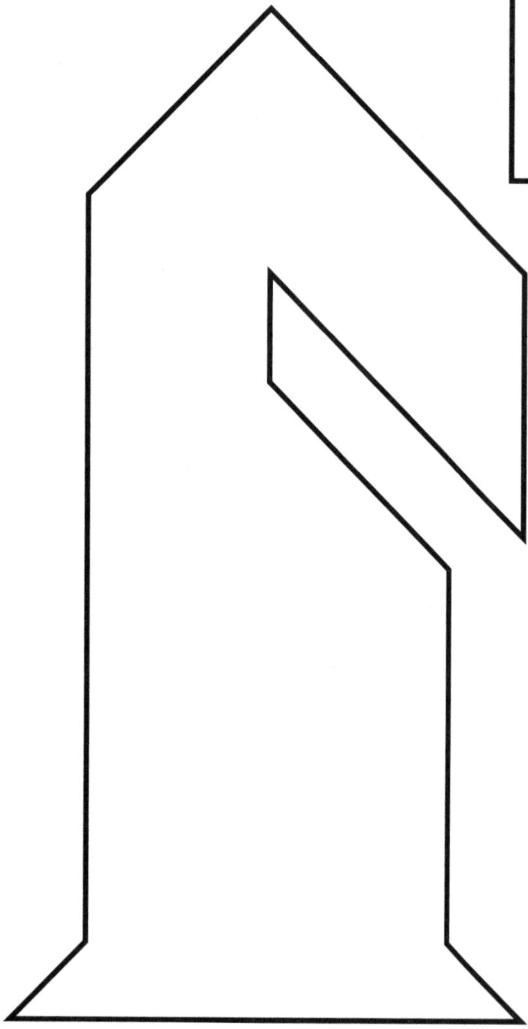

Make this penguin using five shapes.

Show how you did it.

Penguin

SHAPES, SOLIDS, AND MORE

Animals Take Shape

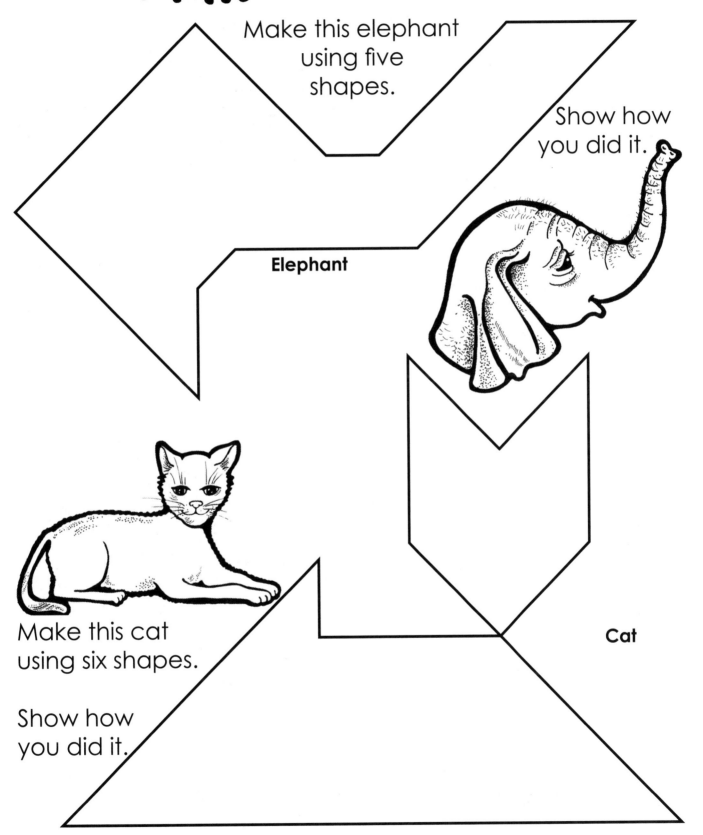

Make this elephant using five shapes.

Show how you did it.

Elephant

Make this cat using six shapes.

Show how you did it.

Cat

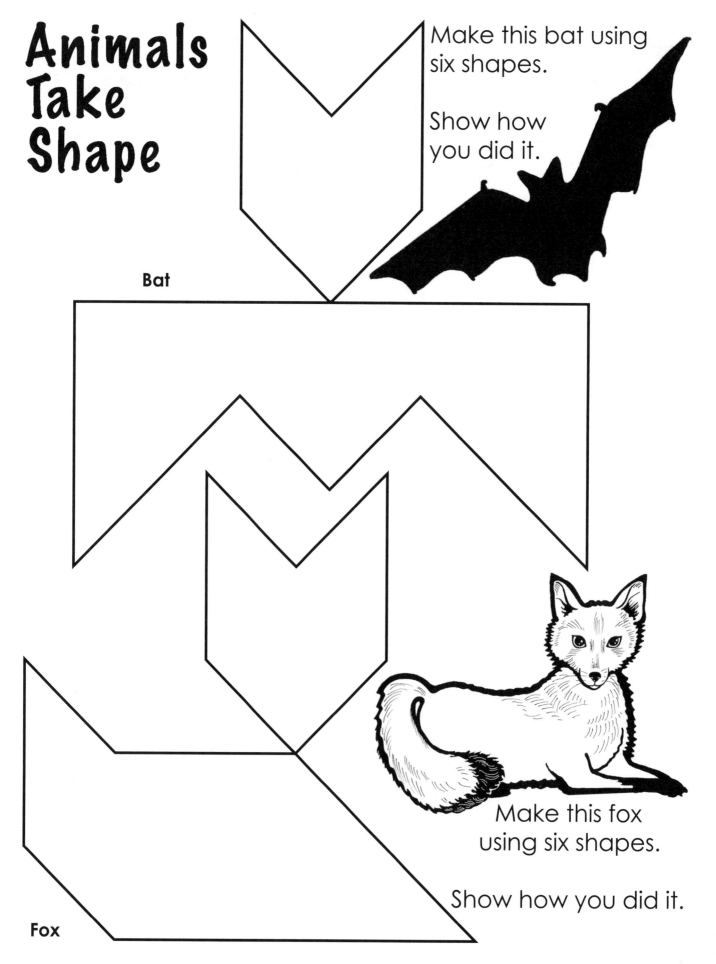

Animals Take Shape

Make this bird using all seven shapes.

Show how you did it.

Bird

SHAPES, SOLIDS, AND MORE © 2009 AIMS Education Foundation

Animals Take Shape

Make this goose using all seven shapes.

Show how you did it.

Goose

SHAPES, SOLIDS, AND MORE

Animals Take Shape

Connecting Learning

1. What shapes are your puzzle pieces?

2. What do you notice about all of the triangle pieces?

3. What else do you notice about the puzzle piece shapes?

4. What shapes could you make by combining the two small triangles?

5. What shapes could you make by combining the three smallest triangles?

Animals Take Shape

Connecting Learning

6. How did you decide which puzzle pieces to use for each puzzle? Could you ever use different pieces to solve the same puzzle?

7. Which puzzle was the easiest for you to solve? Why?

8. Which puzzle was the most difficult for you to solve? Why?

Chart the Parts

Topic
Shape decomposition and identification

Key Question
What hidden shapes can you find in these shapes?

Learning Goals
Students will:
- divide squares, rectangles, triangles, trapezoids, parallelograms, and rhombuses using one fold; and
- identify the smaller shapes into which these shapes can be divided.

Guiding Document
NCTM Standards 2000*
- *Investigate and predict the results of putting together and taking apart two- and three-dimensional shapes*
- *Create mental images of geometric shapes using spatial memory and spatial visualization*

Math
Geometry
    characteristics of 2-D shapes
    shape decomposition
    shape identification
    spatial sense

Integrated Processes
Observing
Collecting and recording data
Organizing data
Comparing and contrasting

Materials
Scissors
Glue sticks
Paper shapes, 3-4 each per student
    (see *Management 1*)
Recording sheets, 6 per group (see *Management 2*)

Background Information
A great deal of emphasis has been placed on the concept of shape composition—creating larger shapes by putting several smaller shapes together. This activity provides an opportunity for students to look at shape decomposition—breaking up larger shapes into their smaller component parts. While an activity on decomposition may seem redundant given all of the time spent on composition, it is important for primary students to have experiences that require them to break shapes down as well as build them up. This helps to reinforce properties and characteristics of shapes in their minds. It also facilitates their understanding that the shapes into which triangles, rectangles, squares, etc., can be broken down are the same shapes that can be used to recreate them.

Management
1. The easiest way to make the shapes necessary for this activity is to use a die cut machine to cut multiple copies of each. If you do not have access to such a machine, templates have been provided that you can photocopy and cut out. The shapes should be cut out of colored paper so that they contrast with the recording sheets on which students will glue them. Use the shapes that are appropriate for your purpose.
2. Each group will need six 12" x 18" recording sheets on which they will glue their solutions. Each page will be used for a different shape. Have students glue their solutions on the pages so that the two decomposed pieces are positioned very close together so the original shape can still be distinguished.

    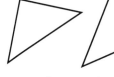

Correct   Incorrect

Procedure
1. Hand out four cut-outs of each shape to students. Identify each one by name.
2. Explain to students that they are going to look for the hidden shapes within these shapes.
3. Show students how to find different hidden shapes by folding the shapes one time in a variety of ways. For example, fold from corner to corner, in half horizontally, in half vertically, etc.
4. Have students try to find at least three different sets of hidden shapes within each of the original six shapes by making one fold only.
5. When all students have found several hidden shapes, explain that they are now going to cut along the fold lines and glue their solutions on the 12" x 18" recording sheets.

SHAPES, SOLIDS, AND MORE

6. Hand out the recording sheets, scissors, and glue sticks to each group. Instruct them to glue an uncut shape in the top left corner of each paper. This shape will let them know which solutions to glue on that page. Have students glue each solution they found on the appropriate recording sheet. If they are able, have students identify the hidden shapes in each solution.

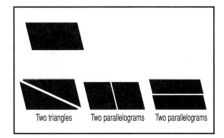

7. Close with a time of class discussion where groups compare their solutions and share what they have learned.

Connecting Learning
1. What hidden shapes were you able to find in the square? ...the rectangle? ...the triangle? ...etc. (See *Solutions* for samples.)
2. Are all of the hidden triangles the same? ...the hidden trapezoids? ...the hidden parallelograms? ...etc. Why or why not?
3. What hidden shapes were not found in the square? ...the rectangle? ...the triangle? ...etc.
4. Are any of those shapes possible using different folds? [Students may not have discovered all of the possible shapes that can be made using one fold. If it is appropriate, you may wish to guide the class into a group discovery of additional hidden shapes.] Why or why not?

Extension
Have students determine what shapes the square, rectangle, triangle, etc., can be divided into using two folds instead of just one. Be sure to open the shapes after the first fold before making the second fold. (See *Solutions* for examples.)

Solutions
A few examples of the common folds are shown here.

Parallelograms—One Fold

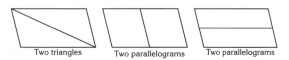

Rectangles—One Fold

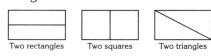

Rhombuses—One Fold

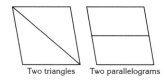

Squares—One Fold

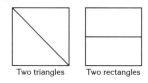

Trapezoids—One Fold

Triangles—One Fold

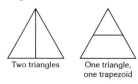

Extension
Samples of the hidden shapes that can be created with two folds are shown for each of the six figures. There are many other possibilities not given here.

Parallelograms—Two Folds

Rectangles—Two Folds

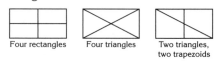

Rhombuses—Two Folds

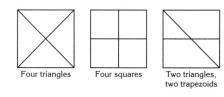

Squares—Two Folds

Four triangles | Four squares | Two triangles, two trapezoids

Trapezoids—Two Folds

Two parallelograms, two trapezoids | Four triangles | Two triangles, one rectangle

Triangles—Two Folds

Three triangles, one quadrilateral | Two triangles, two trapezoids | Two triangles, one pentagon

* Reprinted with permission from *Principles and Standards for School Mathematics*, 2000 by the National Council of Teachers of Mathematics. All rights reserved.

Chart the Parts

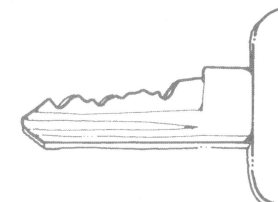

Key Question

What hidden shapes can you find in these shapes?

Learning Goals

Students will:

- divide squares, rectangles, triangles, trapezoids, parallelograms, and rhombuses using one fold; and

- identify the smaller shapes into which these shapes can be divided.

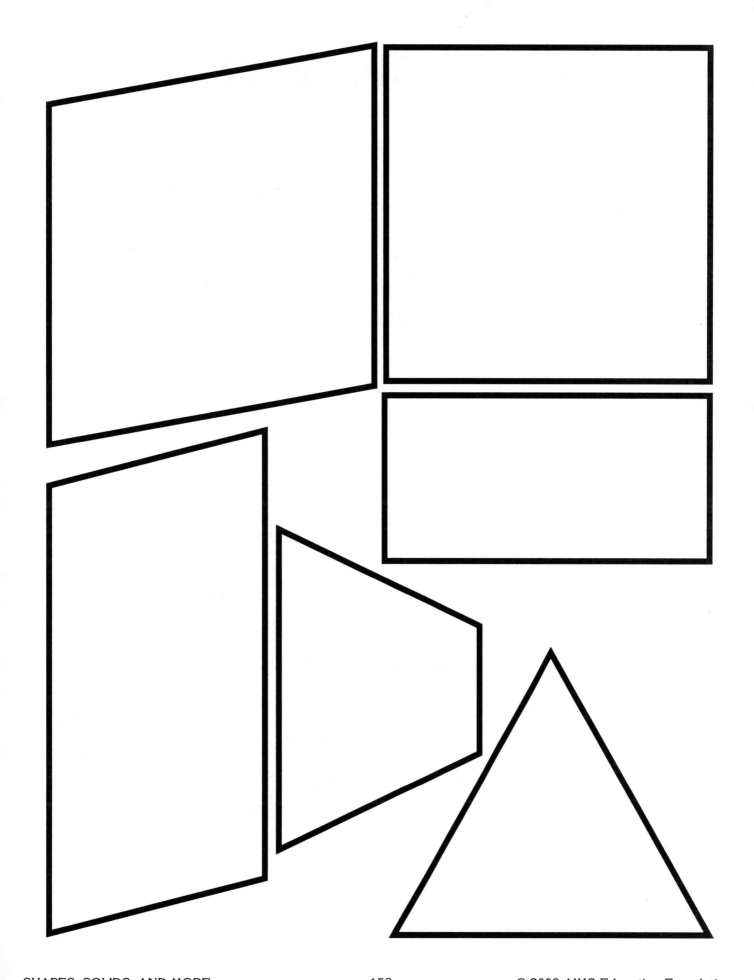

Chart the Parts

Connecting Learning

1. What hidden shapes were you able to find in the square? ...the rectangle? ...the triangle? ...etc.

2. Are all of the hidden triangles the same? ...the hidden trapezoids? ...the hidden parallelograms? ...etc. Why or why not?

3. What hidden shapes were not found in the square? ...the rectangle? ...the triangle? ...etc.

4. Are any of those shapes possible using different folds? Why or why not?

Geometric Gallery

Topic
Geometry

Key Question
How many different geometric elements can you find in pictures?

Learning Goal
Students will identify examples of geometric elements in pictures.

Guiding Documents
Project 2061 Benchmarks
- *Numbers and shapes—and operations on them help to describe and predict things about the world around us.*
- *Shapes such as circles, squares, and triangles can be used to describe many things that can be seen.*
- *Circles, squares, triangles, and other shapes can be found in nature and in things people build.*

*NCTM Standards 2000**
- *Identify, compare, and analyze attributes of two- and three-dimensional shapes and develop vocabulary to describe the attributes*
- *Recognize geometric shapes and structures in the environment and specify their location*

Math
Geometry
    2-D shapes
    3-D objects

Integrated Processes
Observing
Identifying
Communicating

Materials
Pictures (see *Management 1*)

Background Information
    Our world is full of shapes, lines, and patterns: geometry. From the line of a tree branch that looks like a ray as it reaches upward from the trunk, to the cylinders that hold our soups and vegetables, geometry is everywhere. Through the vocabulary of geometry and close observation, we can communicate our understanding of the world in which we live.
    In this activity, students will be asked to observe several pictures and communicate what they see through geometric vocabulary terms. As the students find different geometric elements in the pictures, they will use a wide variety of geometric vocabulary in just a few minutes.

Management
1. Cut out pictures from magazines or newspapers or use your own photos. Select photos with at least one easily identifiable geometric element. A couple of pictures have been included in this activity for those who have trouble locating good examples.
2. Make a transparency of one picture, and mount the others on sheets of construction paper to look like matted pictures. Place the matted pictures around the room as if displayed in an art gallery. Place a piece of paper below each picture for students to record the elements they observe.
3. Students should have many and varied experiences in identifying shapes prior to doing this activity.
4. You may want to make a word bank of geometric terms on the board or bulletin board.

Procedure
1. Tell the class that they will be looking at a series of pictures through the eyes of a mathematician. Explain that you would like for them to find examples of various lines, shapes, and other geometric elements in the pictures.
2. Display the first picture on the overhead. Ask the students to simply look at the picture for 30 seconds without talking. After 30 seconds, invite volunteers to come forward and identify any geometric elements they see in the picture.
3. Direct students' attention to the geometric gallery. Rotate small groups of students through each picture. Ask students to observe the pictures closely and to record any geometric elements they see.
4. End with a discussion about what elements they found in each picture and where else they might find geometric elements in the real world.

SHAPES, SOLIDS, AND MORE © 2009 AIMS Education Foundation

Connecting Learning
1. How many different geometric elements did you see in the first picture? What were they?
2. Were there any geometric elements that appeared in all pictures? What were they?
3. What elements were the hardest to find? ...easiest?
4. Other than in photos, where might we see geometric elements?

Extension
Have students draw a picture and include as many geometric elements as possible.

* Reprinted with permission from *Principles and Standards for School Mathematics*, 2000 by the National Council of Teachers of Mathematics. All rights reserved.

Geometric Gallery

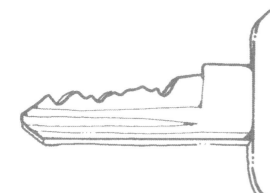

Key Question

How many different geometric elements can you find in pictures?

Learning Goal

identify examples of geometric elements in pictures.

Geometric Gallery

Connecting Learning

1. How many different geometric elements did you see in the first picture? What were they?

2. Were there any geometric elements that appeared in all pictures? What were they?

3. What elements were the hardest to find? …easiest?

4. Other than in photos, where might we see geometric elements?

SHAPES ALL AROUND US

Topic
Geometric shapes and lines

Key Question
How many different geometric shapes and lines can you find as we walk around the neighborhood?

Learning Goal
Students will find and draw examples of geometric shapes and lines (parallel, perpendicular, intersecting, and curved) that they see in the real world.

Guiding Documents
Project 2061 Benchmarks
- Circles, squares, triangles, and other shapes can be found in things in nature and in things that people build.
- Shapes such as circles, squares, and triangles can be used to describe many things that can be seen.
- Numbers and shapes—and operations on them—help to describe and predict things about the world around us.
- Many objects can be described in terms of simple plane figures and solids. Shapes can be compared in terms of concepts such as parallel and perpendicular, congruence and similarity, and symmetry. Symmetry can be found by reflection, turns, or slides.

NCTM Standards 2000*
- Recognize geometric shapes and structures in the environment and specify their location
- Build and draw geometric objects

Math
Geometry
   2-D shapes
   lines
      parallel, perpendicular, intersecting, curved

Integrated Processes
Observing
Classifying
Collecting and recording data
Comparing and contrasting

Materials
Clipboards or books to support paper while drawing
Crayons or colored pencils
Chart paper
Student pages

Background Information
Geometric shapes are everywhere, from the foods we eat to the buildings in which we live and work. Shapes surround us at school, in the neighborhood, on the horizon, and in space.

Students can gain an awareness and appreciation of the geometry in our world through observation. The more focused their observations become, the more details they notice. They then become able to connect the language of geometry to all sorts of real-world settings. This activity will provide them the opportunity to recognize shapes and different kinds of lines in the world around them.

Management
1. A walk in the surrounding neighborhood is suggested, but it could also be done around the school grounds.
2. The different pages can be done on different days, with students focusing on a few shapes or lines each time, or all three pages can be distributed for one search.
3. It is assumed that students are already familiar with all of the geometric vocabulary used on the student pages.

Procedure
1. Explain that the class will be taking a geometry walk. They will be looking for lines and shapes found in nature and in objects made by people.
2. Give students the desired activity page(s) and a firm writing surface such as a clipboard or book. Have them look at the shapes and/or lines for which they will be searching on the walk.
3. Guide the walk, stopping frequently to give students time to draw and label their examples.
4. Return to the classroom, and have students share what they found.
5. Create a class list on chart paper for each of the shapes and lines on which you record all of the different examples that students noted. Continue to add to these lists throughout the year as students make additional observations.

Connecting Learning
1. What shapes/lines were hardest to find? …easiest to find?
2. How many different objects did we find that have squares? …triangles? …parallel lines? etc.
3. Look around the classroom. What other things can we add to our lists?
4. Which of the things on our lists are natural? Which are made by people?

Curriculum Correlation
Literature
Hoban, Tana. *Shapes, Shapes, Shapes.* HarperTrophy. New York. 1996.
Color photographs show familiar objects to introduce students to shapes in the environment.

Hoban, Tana. *So Many Circles, So Many Squares.* Greenwillow Books. New York. 1998.
This book explores circles and squares through photographs of wheels, signs, and other familiar objects.

Hoban, Tana. *Spirals, Curves, Fanshapes and Lines.* Greenwillow Books. New York. 1992.
This book has colorful photographs of the geometry in our world.

Art
1. Make a picture using only geometric shapes.
2. Create a collage using one shape of various sizes and colors, or use a combination of one line and one shape.

Home Link
Have students do shape hunts around their homes.

* Reprinted with permission from *Principles and Standards for School Mathematics*, 2000 by the National Council of Teachers of Mathematics. All rights reserved.

SHAPES ALL AROUND US

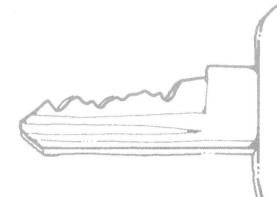

Key Question

How many different geometric shapes and lines can you find as we walk around the neighborhood?

Learning Goal

find and draw examples of geometric shapes and lines (parallel, perpendicular, intersecting, and curved) that they see in the real world.

SHAPES ALL AROUND US

Look for these shapes on your geometry walk. Draw or write the name of each object you find in the right section.

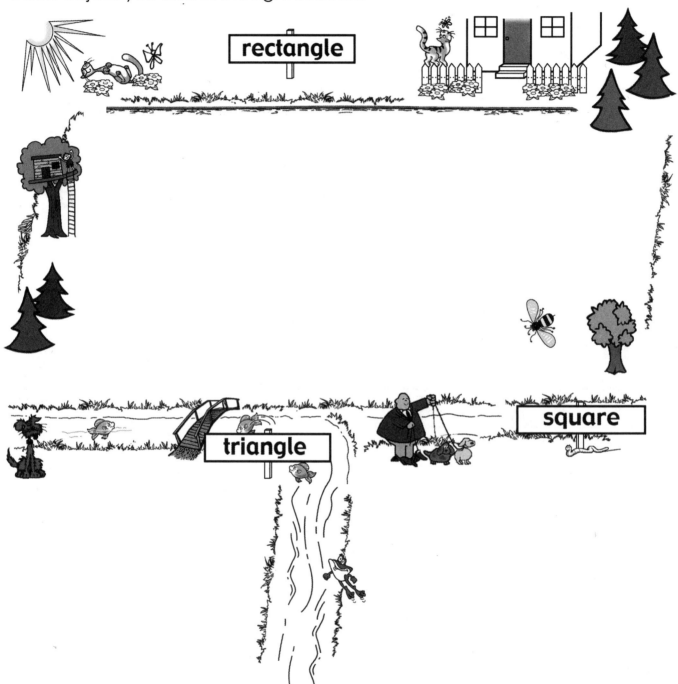

SHAPES, SOLIDS, AND MORE

SHAPES ALL AROUND US

Look for these shapes on your geometry walk. Draw or write the name of each object you find in the right section.

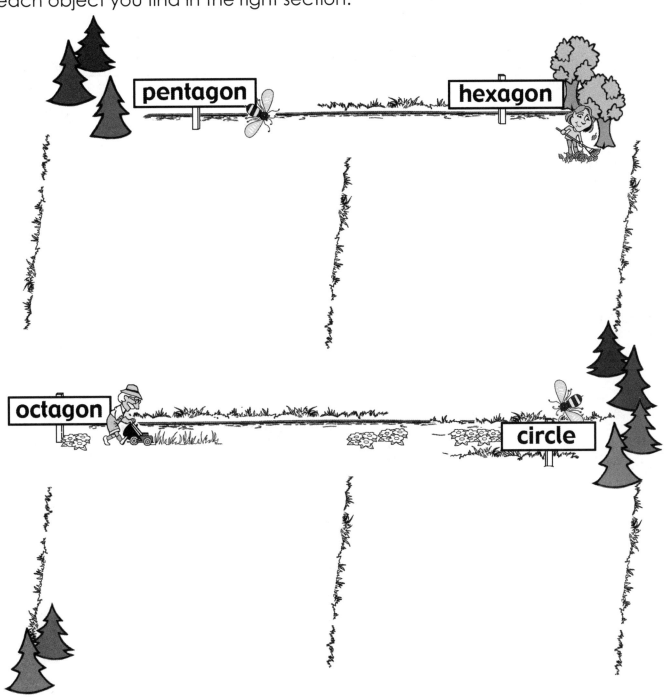

SHAPES, SOLIDS, AND MORE © 2009 AIMS Education Foundation

SHAPES ALL AROUND US

Look for these lines on your geometry walk. Draw or write the name of each object you find in the right section.

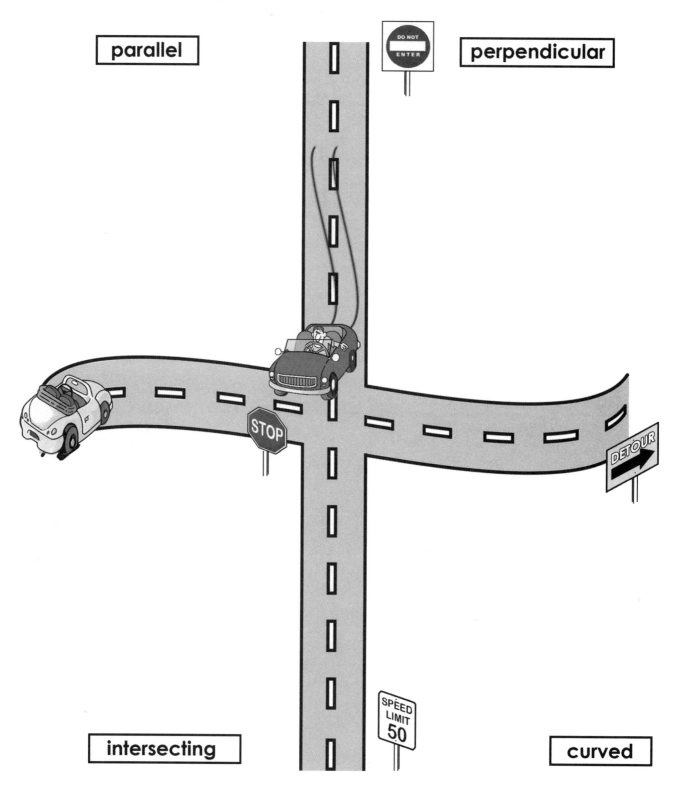

SHAPES ALL AROUND US

Connecting Learning

1. What shapes/lines were hardest to find? …easiest to find?

2. How many different objects did we find that have squares? …triangles? …parallel lines? …etc.?

3. Look around the classroom. What other things can we add to our lists?

4. Which of the things on our lists are natural? Which are made by people?

RADIUS ROUNDUP

Topic
Radius

Key Question
What is the radius of a circle?

Learning Goals
Students will:
- identify the radiuses of circles, and
- determine the lengths of their radiuses.

Guiding Document
NCTM Standards 2000*
- *Describe attributes and parts of two- and three-dimensional shapes*
- *Relate ideas in geometry to ideas in number and measurement*
- *Measure with multiple copies of units of the same size, such as paper clips laid end to end*

Math
Measurement
Geometry
   circles
      radius

Integrated Processes
Observing
Comparing and contrasting
Drawing conclusions

Materials
Small circular candy (see *Management 1*)
Student pages
Paper plates (see *Management 2*)
White glue
Toothpicks
Construction paper
Centimeter rulers

Background Information
    Generally, circle concepts are introduced in third grade and more fully developed in later grades. In this activity, the idea of radius will be addressed. Students should come away from this experience knowing that the *radius* of a circle is the distance from the center of a circle to any point on the circumference of the circle.

Management
1. It is suggested that students use little disc candies (Smarties®) that come wrapped 12-15 to a sleeve as the unit of measure. These can be purchased in large quantity bags and each sleeve will provide students with enough candies to determine the radiuses of all of the circles suggested in this activity.
2. Each pair of students will need one nine-inch and one five-inch paper plate. Be sure the plates are thin and relatively flat so that students will be able to easily fold them.
3. Prior to teaching *Part Two* of this lesson, fold two same-sized paper plates into eighths and darken the radiuses and the centers with a marker.

Procedure
Part One
1. Ask the class if they know what is meant when we talk about the radius of a circle. Explain that it is a straight line from the center of a circle to the edge or circumference of the circle.
2. Give each pair of students one large and one small paper plate. Ask the class to describe the shape of the plates. Question the students about how they would go about finding the radius of each circle. [They first need to locate the center of the circle.]
3. Suggest that the partners each take a plate and that they fold it in half. Once the plates are folded in half, instruct students to fold the plates in half again, thus dividing them into fourths.
4. Have the students mark the point on their plates where all of lines intersect. Explain that this marks the center of the circle. Ask the class how they could find the radius now that the center has been located. [Measure the length of the fold line from the center of the circle to the edge.]
5. Explain that various units could be used to measure the length of the radius such as centimeters, inches, etc. Give each student a sleeve of circular

SHAPES, SOLIDS, AND MORE   © 2009 AIMS Education Foundation

candies and tell them that they will be using candy units to measure the radiuses of their plates. Ask students to measure and record the length of the radiuses of their circles. Compare results.
6. Ask students what they noticed about each of the radiuses. [They were all the same.]
7. End with a discussion in which students share what they have discovered about finding the radiuses of the two plates.

Part Two
1. Ask the class how to find the radius of a circle.
2. Show the class the two paper plates suggested in *Management 3*. Place them near each other on the board and ask the class what they remind them of. [bicycle wheels] Draw the seat and handle bars in if necessary for students to see the connection between the radius of a circle and spokes on a bike.
3. Ask the class how they would measure the radius of a bicycle wheel. [measure the spokes from the center to the outside edge of the tire]
4. Give each student the bicycle page. Have the students find the radius of each circle in centimeter units.
5. Give each student a piece of construction paper and several toothpicks. Ask them to create a wheel using the toothpicks as the spokes. Discuss what the radius and diameter of each circle will be and how they know that. Have the students glue the toothpicks down and draw around the "spokes" to create a unicycle, bicycle, tricycle, etc.
6. Have students measure the radius to the nearest centimeter.
7. End with a discussion that includes other real-world examples of circles and radiuses.

Connecting Learning
1. What is the radius of a circle? [a straight line from the center point of a circle to the circumference, or edge] How do you find the radius of a circle?
2. Why would all of the radiuses for a particular circle be the same?
3. Describe some real-world examples of radiuses. [spokes on a bicycle, pie slices, pizza slices, etc.]

* Reprinted with permission from *Principles and Standards for School Mathematics,* 2000 by the National Council of Teachers of Mathematics. All rights reserved.

RADIUS ROUNDUP

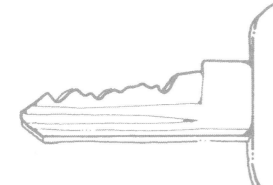

Key Question

What is the radius of a circle?

Learning Goals

Students will:

- identify the radiuses of circles, and
- determine the lengths of their radiuses.

RADIUS ROUNDUP

Measure the radiuses found on the bicycles.

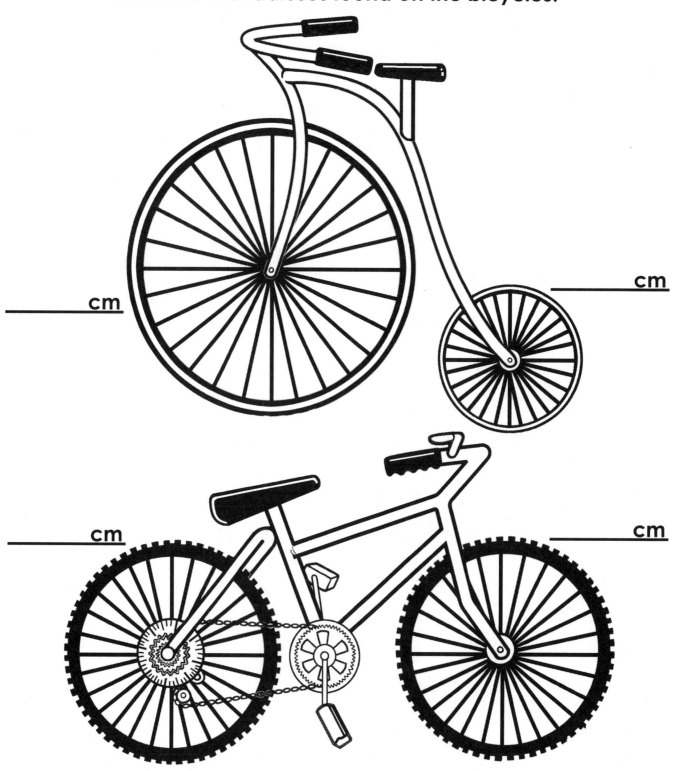

RADIUS ROUNDUP

Connecting Learning

1. What is the radius of a circle? How do you find the radius of a circle?

2. Why would all of the radiuses for a particular circle be the same?

3. Describe some real-world examples of radiuses.

Delivering Diameters

Topic
Diameter

Key Questions
1. How big is nine-inch pizza?
2. What does the size of a pizza have to do with the diameter of a circle?

Learning Goals
Students will:
- find the diameters of several circles, and
- make connections between the diameter of a circle and real-world applications.

Guiding Documents
Project 2061 Benchmark
- When people care about what is being counted or measured, it is important for them to say what the units are (three degrees Fahrenheit is different from three centimeters, three miles from three miles per hour).

NCTM Standards 2000*
- *Describe attributes and parts of two- and three-dimensional shapes*
- *Relate ideas in geometry to ideas in number and measurement*
- *Understand such attributes as length, area, weight, volume, and size of angle and select the appropriate type of unit for measuring each attribute*
- *Use tools to measure*

Math
Measurement
Geometry
   circles
      diameter

Integrated Processes
Observing
Comparing and contrasting
Drawing conclusions

Materials
Pizza rounds (see *Management 1*)
Pizza picture (see *Management 2*)
Pizza toppings pictures
Glue sticks
Scissors
Play clay
Rulers

Background Information
    Third grade students are often taught that the distance across a circle through the center is called the *diameter* and that the *radius* of a circle is the distance from the center of a circle to any point on the edge of the circle. They are also taught how to measure the actual diameter and radius in centimeters, inches, etc., but seldom are they given real-world applications for those skills.
    In this activity, students will explore the connection between finding the radius and diameter of a circle and a pizza. For example, when we buy a nine-inch pizza, it means that it has a diameter of nine inches. Once the center is established and more lines are cut, we can look at a slice of pizza and see that from the center point to the crust would be the radius of the pizza. The students will discover that the same ideas apply to any circle.

Management
1. Prior to teaching this lesson, gather various sized cardboard pizza rounds. Often pizza parlors will donate these or they can be purchased at warehouse type stores. Adjust the questions to whatever size pizza rounds are available.
2. Each student will need a copy of the pizza picture.

Procedure
1. Ask the class how big they think a nine-inch pizza would be.
2. Display a nine-inch pizza round. Discuss where the nine-inch measurement comes from. [its diameter]

SHAPES, SOLIDS, AND MORE

3. Locate the center of the pizza round and draw a line from one edge through the center across to the opposite edge. Invite a student to measure the line to the nearest inch. Ask the class what the measurements would be of other lines that go from edge to edge always passing through the center. [All diameters will be the same length.]

4. Give each student a copy of the pizza picture. Ask them how they would go about finding out what size pizza is on the mat. [Find the center, then measure from one edge through the center across to the opposite edge.] Distribute the rulers. Allow time for them to measure and determine the size of the pizza to the nearest whole inch.

5. Divide the class into small groups. Distribute the pizza toppings page, scissors, and glue sticks. Give each group one of the other pizza rounds. Ask students to find out what size pizza would go on their particular round based on the diameter of the cardboard circle. Ask them to identify the diameter by cutting out and gluing that many pieces of the toppings of their choice. For example, if their pizza round is eight inches, they should put eight pepperoni, eight mushrooms, etc., on it.

6. Have groups share their pizza rounds with the rest of the class, letting the others determine if the diameters of the pizzas are correctly displayed.

7. Give each student a small amount of play clay. Ask the students to make a pizza with a diameter of four inches. Walk around and ask students to prove that their pizza is truly a four-inch pizza. Repeat this procedure several times, changing the desired size of pizza each time.

8. End the lesson with a discussion about how to determine the diameter of a circle and the connection between the diameter of a circle and pizza.

Connecting Learning
1. What is the diameter of a circle? [It is a straight line that passes through the center point of a circle from one edge to the other edge.]
2. What does the diameter of a circle have to do with pizza?
3. What would the diameter of a 12-inch pizza be? How do you know this without measuring the pie?
4. Describe other real-world connections to the diameter of a circle.

* Reprinted with permission from *Principles and Standards for School Mathematics,* 2000 by the National Council of Teachers of Mathematics. All rights reserved.

Delivering Diameters

Key Questions

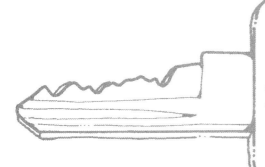

1. How big is nine-inch pizza?
2. What does the size of a pizza have to do with the diameter of a circle?

Learning Goals

Students will:

- find the diameters of several circles, and
- make connections between the diameter of a circle and real-world applications.

Delivering Diameters

What is the size of this pizza?

Delivering Diameters

SHAPES, SOLIDS, AND MORE

Connecting Learning

1. What is the diameter of a circle?

2. What does the diameter of a circle have to do with pizza?

3. What would the diameter of a 12-inch pizza be? How do you know this without measuring the pie?

4. Describe other real-world connections to the diameter of a circle.

All Around Circles

Topic
Circumference

Key Question
What is the circumference of a circle and how can it be measured?

Learning Goals
Students will:
- learn that circumference is the distance around a circle, and
- practice various techniques for measuring the circumference of a circle.

Guiding Document
*NCTM Standards 2000**
- *Describe attributes and parts of two- and three-dimensional shapes*
- *Relate ideas in geometry to ideas in number and measurement*
- *Understand such attributes as length, area, weight, volume, and size of angle and select the appropriate type of unit for measuring each attribute*
- *Understand how to measure using nonstandard and standard units*
- *Use tools to measure*

Math
Geometry
  circles
    circumference
Measurement

Integrated Processes
Observing
Comparing and contrasting
Relating
Applying

Materials
Wind-up meter tape, 30 m
String
Meter sticks
Metric measuring tapes
Assorted objects (see *Management 2*)
Student page

Background Information
Circumference is the distance around a circle. It is equivalent to the perimeter of a polygon, but it has its own name. Because circles are round, directly measuring circumference can be challenging. Many methods can be employed, and some work better than others, depending on the circle being measured. One method is to use a flexible measuring tape to directly measure the circle. This method works well when finding the circumference of jars or cans. Another method is to use a piece of string to go around the circle, then straighten the string and measure it using a ruler or meter stick. This works well with circles drawn on paper or paper plates. A third method that can be used when the circle is made from a flexible material (such as a rubber band or a loop of yarn) is to collapse the circle into a straight line, measure its length, and double it.

Management
1. If you do not have a wind-up meter tape, you will need to tape several metric measuring tapes end to end so that you have 25-30 meters.
2. Gather a variety of circular or cylindrical objects like rubber bands, paper plates, grouping circles, cans, jars, lids, etc., in as many different sizes as possible to give multiple opportunities for finding circumference.
3. You need a large, open area either indoors or outdoors that will allow your entire class to stand in a line, holding hands, with their arms outstretched as far as they can.
4. For *Part Two*, set up a table with the string, meter sticks, and metric measuring tapes. Have enough of each so that several students can be using them at the same time.

Procedure
Part One
1. Ask students what geometry terms they have heard associated with circles. Record their responses on the board. If no on mentions it, write the word *circumference*.
2. Tell students that today they will be learning about circumference. Ask if anyone knows what the circumference of a circle is.
3. Discuss that the circumference is the distance around a circle. It's the same as the perimeter of a square, rectangle, triangle, or other polygon, it just has a unique name that only applies to circles.

SHAPES, SOLIDS, AND MORE   181   © 2009 AIMS Education Foundation

4. Use the end of a can to trace a circle on the board. Challenge students to think about how they could measure the circumference of the circle you just drew.
5. After listening to some of their suggestions, tell students that they are going to try a few different ways to measure circumference. Take the class to the open area you have selected. Instruct them to stand in a circle holding hands. Have the students spread apart so that the circle is as large as it can be without letting go of their hands. (Students' arms should all be fully extended.)
6. Ask students how they could measure the distance around the circle formed by their bodies—the circle's circumference. Show them the wind-up meter tape.
7. Try all of the students' ideas that are feasible and will result in an accurate measure of the circumference, being sure to include the following three methods:
   a. Use the meter tape to measure all the way around the circle. Since the tape is flexible, it can be used to directly measure the circumference.
   b. Turn the circumference into a straight line. Select two adjacent students to let go of their hands and begin to walk apart so that the circle becomes a straight line. Have students continue to keep their arms fully extended so that the distance remains the same. Use the wind-up meter tape to measure the length from the first student's outstretched arm to the last student's outstretched arm.
   c. Have the circle "fold in half." Direct the students to walk toward each other until they are making a line, two people deep while still holding hands and keeping their arms outstretched. Measure the distance from one end of the line to the other, then double that distance to find the circumference.
8. Have students return to the classroom. Discuss the different methods used to find the circumference and how they compared.

Part Two
1. Explain that students are now going to have the opportunity to apply the same methods you used to find the circumference of their circle to their own circles.
2. Distribute a circular object and the student page to each student. Show students the table with the measuring tools that are available. Invite them to decide on one method for finding the circumference of the circle and to select the appropriate measuring tool(s) for this method.
3. Have them record their measurements on the student page along with the method used to find the circumference.

4. Instruct students to trade objects and to repeat the process using a different method of finding the circumference. (Remind them to return the measuring tools to the table once they are finished so that others can use them.)
5. Repeat until all students have used at least three different methods for finding the circumference.
6. Discuss which method students prefer and why.

Connecting Learning
Part One
1. What is the circumference of a circle? [the distance around the circle]
2. What are some ways that circumference can be measured?
3. How did the different ways we measured the circumference of our "student circle" compare? Which do you think was easiest? …most difficult?

Part Two
1. What methods did you use to find the circumferences of your circular objects?
2. How did the circumferences of your circles compare? [the bigger the circle, the larger the circumference]
3. Which method for finding the circumference did you like best? Why?
4. Do you think one method is more accurate than another? Justify your response.
5. What method would you use to find the circumference of a rubber band? Why? [It is difficult to measure around a rubber band with either string or a tape measure, so folding it in half and the doubling the length is likely the easiest way to find the circumference.]
6. When would you need to know the circumference of a circle?

* Reprinted with permission from *Principles and Standards for School Mathematics*, 2000 by the National Council of Teachers of Mathematics. All rights reserved.

All Around Circles

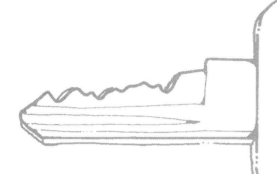

Key Question

What is the circumference of a circle and how can it be measured?

Learning Goals

- learn that circumference is the distance around a circle, and
- practice various techniques for measuring the circumference of a circle.

All Around Circles

"My arms are too short or my circle is too big!"

Describe each circle. Find its circumference. Tell how you found the circumference. Use a different way each time.

My circle: _____ Circumference: _____

How I measured the circumference: _____

_____

My circle: _____ Circumference: _____

How I measured the circumference: _____

_____

My circle: _____ Circumference: _____

How I measured the circumference: _____

_____

My circle: _____ Circumference: _____

How I measured the circumference: _____

_____

SHAPES, SOLIDS, AND MORE © 2009 AIMS Education Foundation

All Around Circles

Connecting Learning

Part One

1. What is the circumference of a circle?

2. What are some ways that circumference can be measured?

3. How did the different ways we measured the circumference of our "student circle" compare? Which do you think was easiest? ...most difficult?

Part Two

1. What methods did you use to find the circumferences of your circular objects?

All Around Circles

Connecting Learning

2. How did the circumferences of your circles compare?

3. Which method for finding the circumference did you like best? Why?

4. Do you think one method is more accurate than another? Justify your response.

5. What method would you use to find the circumference of a rubber band? Why?

6. When would you need to know the circumference of a circle?

Circle Concepts

Topic
Circles

Key Question
Where are the diameter, center, radius, and circumference of a circle found?

Learning Goals
Students will:
- develop an awareness of the properties of a circle—diameter, center, radius, and circumference;
- fold a circle in two directions symmetrically to find its center; and
- find the circumference, diameter, and radius of various circles.

Guiding Document
NCTM Standards 2000*
- *Describe attributes and parts of two- and three-dimensional shapes*
- *Relate ideas in geometry to ideas in number and measurement*
- *Understand such attributes as length, area, weight, volume, and size of angle and select the appropriate type of unit for measuring each attribute*
- *Understand how to measure using nonstandard and standard units*
- *Use tools to measure*

Math
Geometry
  circles
Measurement
Estimation
Equalities and inequalities

Integrated Processes
Observing
Comparing and contrasting
Collecting and recording data
Interpreting data
Drawing conclusions

Materials
For each student:
  tape measures
  24-inch piece of string
  crayons, in three colors
  paper
  round coffee filter
  plastic lid
  pushpin
  permanent marker

Background
Circle concepts are usually introduced in third grade. The concepts, while not difficult, introduce new vocabulary to the students—diameter, center, radius, and circumference. Here are some kid-friendly definitions for those geometry terms:
- diameter—any straight line that goes from one side of the circle to the other and passes through the center
- center—the point that marks the middle of a circle
- radius—any line from the center of a circle to the edge
- circumference—the distance around a circle

In this activity, students will identify the parts of a circle and look at real-world examples of circles, such as pizzas and pies, which will help students think about diameter and radius.

Management
1. Prior to teaching this lesson, iron one coffee filter per student.
2. Gather plastic lids in various sizes so that you have enough for one per student. Use a permanent marker to identify each lid with a letter.

Procedure
Part One
1. Give students paper and ask each one to draw a circle and cut it out.
2. Explain that if they have drawn a perfect circle, they will be able to fold it in half time and time again and the halves will perfectly overlay each other. Have the students fold their circles. Discuss whether they were able to draw a perfect circle and why drawing circles is so hard to do.

SHAPES, SOLIDS, AND MORE 187 © 2009 AIMS Education Foundation

3. Tell the class that for this lesson, you will provide them with a circle. Distribute tape measures and coffee filters. Ask the students to estimate the location of the center of the coffee filter and use a pencil to mark a dot there. Discuss how they could test to see that the point they made is actually in the center of the circle. [They could fold the circle in half and then into fourths and the point where the lines intersect would be the center. They could also measure from the dot to the edge at various points around the edge of the circle and each radius measurement should be the same length.]
4. Have the students fold their coffee filter circles in half so that the halves overlay each other. Instruct students to open the coffee filters and to use a crayon to mark the fold line. Explain that the diameter of a circle is the length of a straight line that passes through the center of the circle and ends at the circle's edge. Have students label the crayon line *diameter* and find its length.
5. Ask students to fold their coffee filters in half so that the crayon line (or diameter) overlays itself. (This should make four 90-degree sectors when the filter is opened up.) Have the students unfold the circles and mark the point where the lines intersect with a different colored crayon and label it the *center*. Ask the students how they would describe what a center is to someone who didn't know about circles.
6. Tell students that the *radius* of a circle is any line segment from its center to its edge. Invite the students to identify one radius on their circles by using a third crayon to darken the appropriate line and labeling it. Have students also measure and record the length of the radius.
7. Discuss real-world examples where we find the center, diameter, and radius. [a round pizza or pie that has been cut into equal pieces]

Part Two
1. Review the parts of a circle identified in *Part One* of this lesson. Tell the students that there is one part of a circle that has not yet been discussed and that is the *circumference* of the circle. Explain that the circumference of a circle is the distance around the circle. Discuss ways that the circumference of a circle can be measured. [Wrap a piece of string around the circle, cut it or mark it at the point where the two ends meet, and hold the length of string against a measuring tape.]
2. Give each student a plastic lid, pushpin, piece of string, a measuring tape, and a piece of paper.
3. Ask students to write their names and the letters of their lids on a piece of paper. Invite a student to define the circumference of a circle. Have students work in pairs to find the circumferences of their circles. One student should hold one end of the string while the other student wraps the string around the lid and marks or cuts where the strings meet. Students can then use their measuring tapes to determine the length of the string.
4. Ask a student with a relatively small lid to tell you the circumference of his/her circle. Have a student with a larger lid hold his/her lid up and invite students to predict what the circumference of that lid would be. Discuss whether they think the circumference would be greater than, less than, or equal to the smaller lid. Allow the student with the larger lid to report its circumference.
5. Tell students to trace around their lids on paper, cut them out, then fold the papers to find the centers. Have them place the paper circles on top of their lids and mark the center with a pushpin. When they have established that the pushpins are directly in the centers of their lids, ask the students to find and record both the diameters and radii of their lids. Again, compare two lids in the class and allow students to predict the radius and diameter of one based on what they know about the other.
6. Have students exchange lids. Ask them to find the radius, circumference, and diameters of each and record the information on their papers.
7. Discuss the students' findings and explore other circles in the classroom such as the trashcan base, classroom clock, the lip of a cup or glass, etc.

Connecting Learning
1. What are the main parts of a circle? Where are they found?
2. If I fold a circle in half so that both sides are symmetrical, what does the fold line represent? [diameter]
3. What is the distance around a circle called? [circumference] How would you measure it?
4. How can I tell if I have located the exact center of a circle?
5. How do the radius and diameter of a circle compare? [The radius is half the diameter.]
6. When in the real world do we cut circles on diameter lines? [pies, pizzas, etc.]

* Reprinted with permission from *Principles and Standards for School Mathematics*, 2000 by the National Council of Teachers of Mathematics. All rights reserved.

Circle Concepts

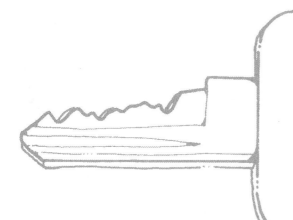

Key Question

Where are the diameter, center, radius, and circumference of a circle found?

Learning Goals

Students will:

- develop an awareness of the properties of a circle—diameter, center, radius, and circumference;
- fold a circle in two directions symmetrically to find its center; and
- find the circumference, diameter, and radius of various circles.

Circle Concepts

Connecting Learning

1. What are the main parts of a circle? Where are they found?

2. If I fold a circle in half so that both sides are symmetrical, what does the fold line represent?

3. What is the distance around a circle called? How would you measure it?

4. How can I tell if I have located the exact center of a circle?

5. How do the radius and diameter of a circle compare?

6. When in the real world do we cut circles on diameter lines?

Suitcase Symmetry

Topic
Symmetry

Key Question
What makes something symmetric?

Learning Goal
Students will examine real-world objects and drawings and determine whether or not they have line symmetry.

Guiding Documents
Project 2061 Benchmark
- *Many objects can be described in terms of simple plane figures and solids. Shapes can be compared in terms of concepts such as parallel and perpendicular, congruence and similarity, and symmetry. Symmetry can be found by reflection, turns, or slides.*

*NCTM Standards 2000**
- *Recognize and create shapes that have symmetry*
- *Identify and describe line and rotational symmetry in two- and three-dimensional shapes and designs*

Math
Geometry
    symmetry

Integrated Processes
Observing
Comparing and contrasting
Classifying

Materials
Various symmetric and nonsymmetric items
    (see *Management 1*)
Glue
Scissors
Paper grocery bag
Masking tape
Overhead transparency (see *Management 5*)
Student pages
Magazines, optional
Mirrors, optional

Background Information
Symmetry is a geometric property of both two- and three-dimensional shapes and solids. While there are many kinds of symmetry, this activity will only address *line symmetry*. An object is said to have *line symmetry* if it can be divided into two equal parts that are mirror images of each other. This line of symmetry, by definition, must pass through the exact center of the object.

Examples of symmetry can be found in nature, art, and architecture. It is important for students to use a variety of materials, such as pattern blocks, paper cutouts, paper folding, and mirrors, to explore lines of symmetry before they are asked to look at a picture and determine whether a figure is symmetric.

One strategy to use when exploring the topic of symmetry with young children is to show the children different objects around the classroom, cover half of each object and ask the child if the hidden half of the object is exactly the same as the half that they can see.

In this activity, we will begin by folding three-dimensional objects to build an understanding of symmetry. We will then progress to both physically and mentally folding two-dimensional shapes and objects to determine whether they are symmetric.

Management
1. Gather several objects and shapes that can be folded to determine whether or not they are symmetric. Items might include: a pair of children's pants, a shirt, a pair of shorts, a hat, a place mat, an irregular piece of paper, a drinking straw, a sock, a slice of bread, a leaf, a re-closable plastic bag, paper plates, etc. Place these items in a large paper grocery bag. When looking for the line of symmetry in items, caution the students to only look at the shape and not the details on the objects such as pockets, zippers, etc.
2. Place a masking tape strip vertically down the center of a child's T-shirt to represent the line of symmetry. Fold the shirt in half vertically so that the tape does not show.
3. When determining whether a two-dimensional shape or picture is symmetric, a mirror can be placed to test the figure or object.

SHAPES, SOLIDS, AND MORE   © 2009 AIMS Education Foundation

4. Magazine pictures can be used instead of, or in addition to, the pictures provided on the student page. Some students may need assistance in cutting out the suitcase and pictures.
5. Copy the mitten and soda can onto an overhead transparency, or enlarge them so students can easily see them.

Procedure
Part One
1. Remove a small child's T-shirt (prepared as directed in *Management 2*) from the paper grocery bag.
2. Ask the class if they think that the hidden half of the shirt is exactly the same as the half they can see. Record their responses on the board.
3. Explain that objects or shapes are called *symmetric* when they can be divided or folded in half so that both sides are the same size and same shape (congruent).
4. Write the word *symmetric* on the board. Read the word aloud to the students.
5. Unfold the T-shirt and ask the students if the shapes of both halves of the shirt are exactly the same. Ask the students if the shirt would be considered symmetric since both halves are the same size and same shape (congruent). [Yes.]
6. Tell the students that the fold line that divides the shape or object in half is called the line of symmetry. Label the tape *line of symmetry*.
7. Ask the students if both halves would be symmetric if you folded the shirt in half so that the shoulders were touching the bottom of the shirt. Invite a student to test his or her thinking by folding the shirt. Discuss the fact that the T-shirt has only one way that it can be folded so that each half is exactly the same shape, therefore it only has one line of symmetry.
8. Tell the students that you have several objects that you would like for them to examine and help you sort as symmetric or nonsymmetric.
9. Display a small child's pair of pants from the grocery bag. Have the students observe the pants, looking for a way to fold the pants so that both halves would be the same size and shape (congruent).
10. Invite a student to come forward and demonstrate how the pants could be folded so that both sides would be the same. Identify where the line of symmetry would be by placing a piece of masking tape on the clothing to represent the line of symmetry.
11. Attempt to fold the pants horizontally. Ask the students if the two sides are symmetric. Guide the students to make the statement that the pants have only one vertical line of symmetry.
12. Place the pants and T-shirt in an area labeled *symmetric*. Label another area *nonsymmetric*.
13. Continue to pull items from the grocery bag, allowing students to examine the items and identify whether or not they are symmetric. Have students place the items in the appropriate areas.
14. Compare the shapes of the items within each group. Guide the students to make some generalizations about the shapes of items that are symmetric. [rectangles, square, round, etc.]

Part Two
1. Ask the *Key Question*.
2. Tell the students that you would like to take them on an imaginary trip to the land of symmetry. Explain that they will be given a "suitcase" and that they will only be allowed to pack things that are symmetric.
3. Review what it means to be symmetric (referring here to line symmetry). [There is a way to divide it into equal parts that are mirror images of each other. The fold line is called the line of symmetry.]
4. Display the picture of the soda can. Ask the students if they could pack the soda in their "suitcase" since they can only take symmetric things. Tell them to focus only on the shape of the can.
5. Draw first a vertical and then a horizontal line on the can in an attempt to divide it symmetrically. Again ask the students if they could pack the soda can. [Yes, because it is symmetric.] Repeat the process with the picture of the mitten. [The picture of the mitten is not symmetric.]
6. Distribute the student pages. Assist any students who may need help in cutting out the suitcase and pictures.
7. Ask the students to look at each picture and decide if the picture of the item is symmetric. Remind them that they can draw lines on the pictures to help them decide.
8. Instruct the students to fold their suitcases so that the illustrations are on the outside. Allow them to color and decorate as desired. Ask students to glue the pictures of all of the items that are symmetric on the inside of the suitcases.
9. When the students have finished the assignment, discuss what items they packed and why certain items could not be taken.

Connecting Learning
1. Explain in your own words what symmetric means.
2. How did you decide if the objects from the bag were symmetric?
3. How would you teach someone else about things that are symmetric?
4. Name some shapes that are symmetric.

5. What do you notice about all of the things that are symmetric? [They can all be divided so that both sides are mirror images of each other.]

Extensions
1. Walk around the schoolyard looking for things in nature that have symmetry: leaves, flowers, insects, etc.
2. Have the students look through illustrations in picture books and identify things that have symmetry.

Curriculum Correlation
Murphy, Stuart J. *Let's Fly a Kite*. HarperCollins. New York. 2000.

Solutions
Symmetric objects are: jeans, shirt, comb, skirt, ball, book (horizontal line of symmetry), boots

* Reprinted with permission from *Principles and Standards for School Mathematics,* 2000 by the National Council of Teachers of Mathematics. All rights reserved.

Suitcase Symmetry

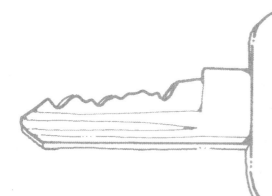

Key Question

What makes something symmetric?

Learning Goal

Students will:

examine real-world objects and drawings and determine whether or not they are symmetric.

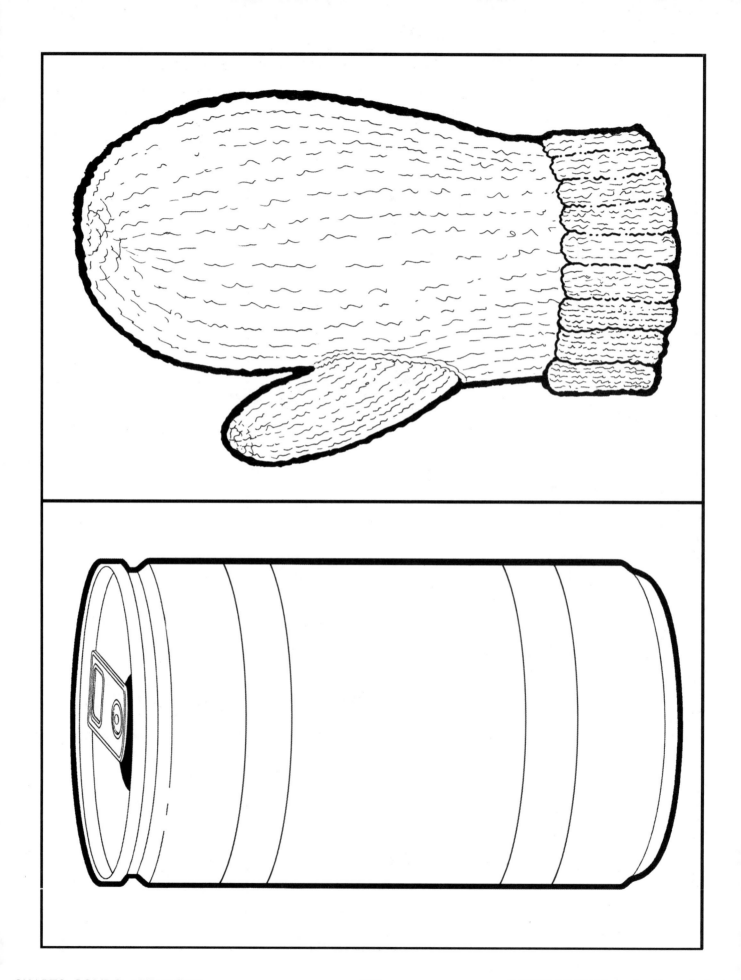

SHAPES, SOLIDS, AND MORE   196   © 2009 AIMS Education Foundation

Suitcase Symmetry

SHAPES, SOLIDS, AND MORE 197 © 2009 AIMS Education Foundation

Suitcase Symmetry

Connecting Learning

1. Explain in your own words what symmetric means.

2. How did you decide if the objects from the bag were symmetric?

3. How would you teach someone else about things that are symmetric?

4. Name some shapes that are symmetric.

5. What do you notice about all of the things that are symmetric?

Sticking to Symmetry

Topic
Symmetry

Key Question
How can you determine the line(s) of symmetry on various shapes?

Learning Goals
Students will:
- explore a variety of symmetric shapes, and
- determine all the lines of symmetry on several geometric shapes.

Guiding Documents
Project 2061 Benchmark
- Many objects can be described in terms of simple plane figures and solids. Shapes can be compared in terms of concepts such as parallel and perpendicular, congruence and similarity, and symmetry. Symmetry can be found by reflection, turns, or slides.

NCTM Standards 2000
- Recognize and create shapes that have symmetry
- Identify and describe line and rotational symmetry in two- and three-dimensional shapes and designs
- Investigate and predict the results of putting together and taking apart two- and three-dimensional shapes

Math
Geometry
    2-D shapes
    symmetry

Integrated Processes
Observing
Comparing and contrasting
Analyzing

Materials
Coffee stirrers (see *Management 1*)
Shapes (see *Management 2*)
Transparent tape
Scissors
Scratch paper
Student page

Background Information
When an object or shape has line symmetry, it can be divided into two halves that are mirror images of each other. The line along which it is divided is the line of symmetry. This is sometimes referred to as the mirror line. Depending on the shape, there may be multiple lines of symmetry. This activity will give students the opportunity to create symmetric shapes and to find the lines of symmetry within shapes.

Management
1. Be sure to get the round coffee stirrers that are like miniature straws, not the flat plastic or wooden kind. Each student needs nine coffee stirrers.
2. Copy the pages of shapes onto card stock. (There is a different page of shapes for each part of the activity.) Each student needs one set of shapes for *Part One*. Pairs of students can share a page of shapes for *Part Two*.
3. This activity is divided into two parts and should be done over the course of two days.
4. It is assumed that students are already familiar with line symmetry and are able to identify objects, shapes, patterns, etc., that exhibit line (mirror) symmetry.

Procedure
Part One
1. Ask students how to tell if something is symmetric. Review the meaning of symmetry and how to find the line of symmetry for a shape.
2. Tell students that they will be exploring some shapes that have symmetry and will begin by making some "symmetry sticks" using coffee stirrers and shapes.
3. Distribute a set of shapes and scissors to each student. Instruct them to carefully cut out each shape.
4. Discuss the shapes that students cut out. What do they seem to be? [half of a star, half of a heart, half of a tree]
5. Give students three coffee stirrers each and transparent tape. Show them how to tape a card stock shape to the coffee stirrer so that the straight edge of the shape aligns with the stirrer. Be sure that students tape the shapes to the tops of the stirrers in an upright position.

SHAPES, SOLIDS, AND MORE   199   © 2009 AIMS Education Foundation

6. When students have taped their shapes to their stirrers, have them lay the stirrers on a piece of scratch paper and trace around the shape and the coffee stirrer. Instruct them to flip the shape while keeping the coffee stirrer in the same place and trace the outline of the shape again. (They can use a finger to flip the shape back and forth on top of the outline and see both halves in quick succession.)

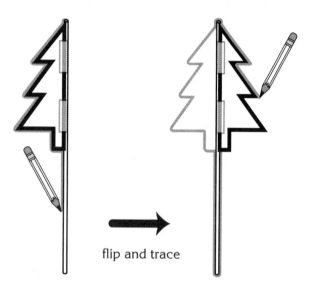

flip and trace

7. Discuss whether or not the shapes students traced have symmetry, and why. Have students identify the line of symmetry in the shapes. [The coffee stirrer serves as the line of symmetry.]
8. Challenge students to think about what the result would have been if they had not taped the shapes to the coffee stirrer along the straight side. [The shape would still have been symmetrical, but it would not have been a heart (star, tree); it would have been something else.]
9. Make a stirrer using one of the shapes where the shape is not taped along the straight side. Trace the shape and its flip on the overhead and have students describe what they see. Discuss how the shape is still symmetric, but it is not the same as the heart (star, tree) that you would see if you taped on the straight side. For these shapes, there is only one way to tape the stirrer so that the shape you create by flipping will be a heart (star, tree).

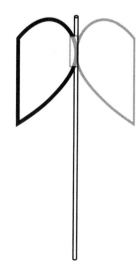

10. If desired, show students how to hold a stirrer between their fingers and spin it by rapidly rubbing their hands back and forth. The effect is enhanced if the stirrer is spun in front of something that is dark in color so that the white of the shape is more evident. Persistence of vision will cause it to appear as if the two halves of the shape are there, and students may even be able to see a three-dimensional representation of the shape. (In three-dimensional space, the coffee stirrer becomes an axis of rotational symmetry rather than being a line of mirror symmetry.)

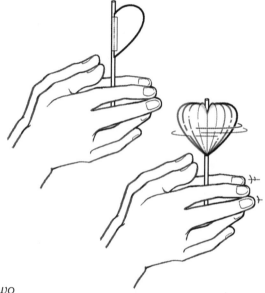

Part Two
1. Remind students of the procedure from *Part One*. Explain that students will be making additional symmetry sticks, but instead of starting with half of a shape, they will be starting with whole shapes. Their challenge will be to decide where to cut the shape and how to tape it to the coffee stirrer so that when it is flipped, the outline of the shape they trace is the same as the original shape.
2. Review the shapes that students had in *Part One* and how the coffee stirrer was the line of symmetry. Remind students of the discussion about where to tape the coffee stirrer so that the shape created was a heart, star, or tree.
3. Distribute the student page. Point out the three shapes on the student page. Explain that these shapes are the shapes they need to create when they trace the shapes on the stirrers. Give each pair of students a page of shapes to cut. Tell them that these are the shapes they will be using to make their symmetry sticks.
4. Ask students how they will decide where to cut each shape. [find the line(s) of symmetry] Review how to find a line of symmetry in a shape.
5. Allow students time to cut out the shapes and cut them along the lines of symmetry. Be sure they

know that there may be multiple solutions and that they should try to find all of the possibilities for each shape.
6. When students have their shapes cut, distribute coffee stirrers and tape so that they can assemble their sticks and test them to see if they are correct.
7. When the correct lines of symmetry have been determined for each shape, instruct students to draw in each line on the student page. (There are more shapes provided than there are lines of symmetry, so some will be blank.)
8. Have students share their sticks and the lines of symmetry they found for each shape. Compare and contrast the lines of symmetry and the resulting symmetry sticks for each shape. For example, the hexagon has six lines of symmetry, but only two different sticks because multiple lines of symmetry divide the shape in the same way.

  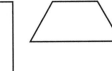

Six lines of symmetry   Two shapes resulting from a cut along a line of symmetry

Connecting Learning

Part One
1. How do you know if something has a line of symmetry? [It can be divided into two halves that are mirror images of each other.]
2. What were the shapes that you made with the symmetry sticks? [heart, star, tree]
3. Are these shapes symmetric? [Yes.] How do you know? [The halves are mirror images of each other.]
4. Where is the line of symmetry in these shapes? [coffee stirrer]
5. What would happen if you taped the shape to the stirrer in a different place? [The resulting shape would be different, but symmetric.]
6. Would the shape you trace be symmetric? How do you know? [Yes; even though it looks different than the original shape, both halves are still mirror images of each other, so it still has symmetry.]

Part Two
1. How do you find the line of symmetry in a shape? [Find a place where it can be divided into two halves that are mirror images of each other.]
2. Can a shape have more than one line of symmetry? [Yes.] How do you know? [If there is more than one way to divide a shape into two halves that are mirror images of each other, it has more than one line of symmetry.]

3. How did you find the lines of symmetry in the shapes you had to cut?
4. How many lines of symmetry did you find in each shape? (See *Solutions*.)
5. Did cutting along each of these lines always give you a different result? [No. For the hexagon and the square, there are multiple ways to cut it that will result in the same two shapes.]
6. What did you learn about symmetry from this activity?

Extensions
1. Give students other geometric shapes and challenge them to create symmetry sticks.
2. Tape each shape from *Part Two* to the coffee stirrer in as many ways as possible. Record all of the shapes that result from flipping and tracing the stirrer.
3. Use mirrors placed along the line of symmetry to view the figures.

Solutions

Hexagon
Six lines of symmetry total. Two unique shapes.

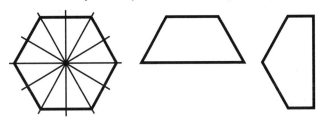

Square
Four lines of symmetry total. Two unique shapes.

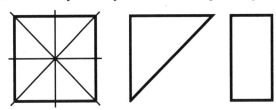

Rhombus
Two lines of symmetry total. Two unique shapes.

* Reprinted with permission from *Principles and Standards for School Mathematics*, 2000 by the National Council of Teachers of Mathematics. All rights reserved.

Sticking to Symmetry

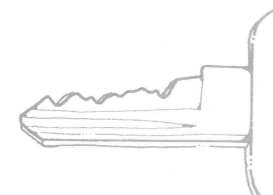

Key Question

How can you determine the line(s) of symmetry on various shapes?

Learning Goals

Students will:

- explore a variety of symmetric shapes, and
- determine all the lines of symmetry on several geometric shapes.

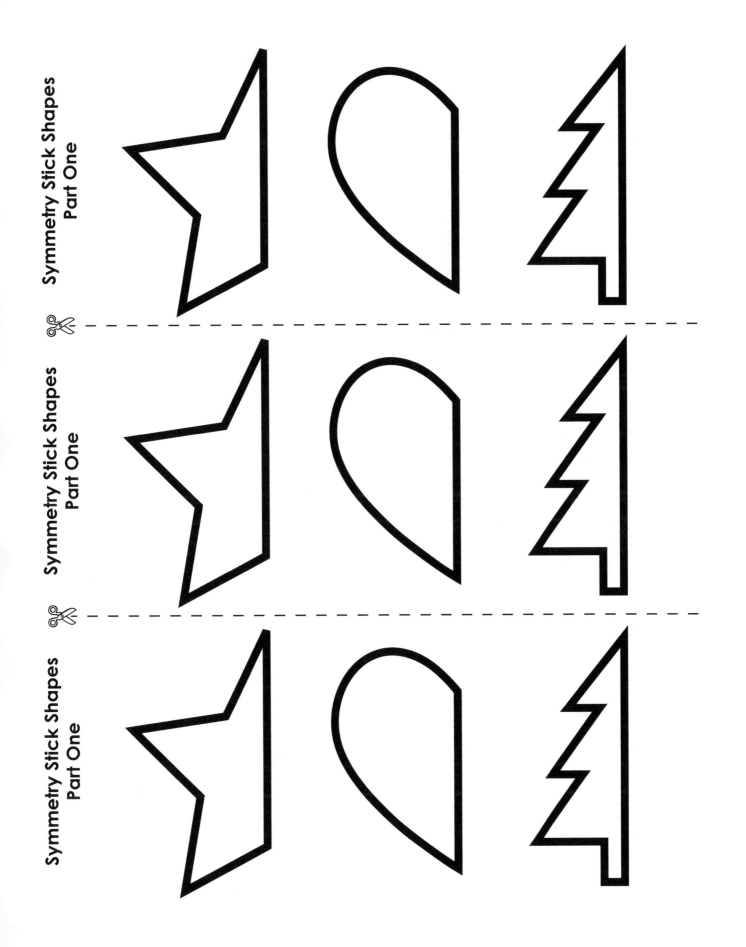

Part Two

Your challenge: Make symmetry sticks that show each of these shapes.

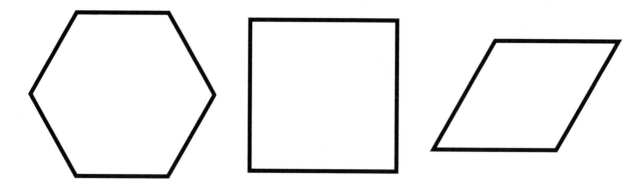

Do this:

1. Cut out one shape from the next page.

2. Decide how to cut the shape. (Find a line of symmetry.)

3. Cut along the line of symmetry.

4. Tape the cut shape to a coffee stirrer. Trace the shape.

5. Flip the stirrer and trace again to see if you got it right.

6. Repeat until you find all the ways to cut each shape.

7. Record each line of symmetry by drawing on the shapes below.

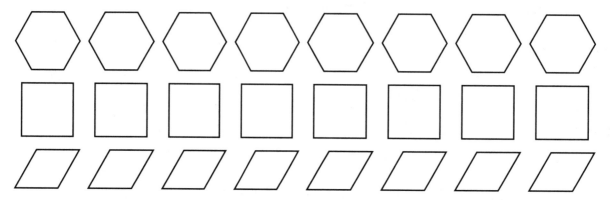

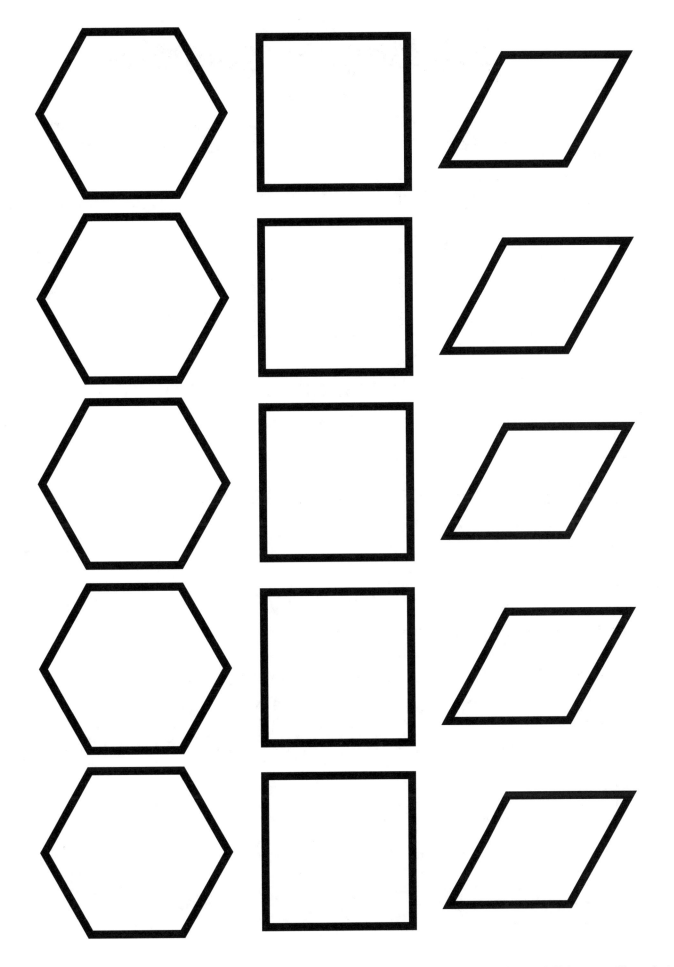

Connecting Learning

Part One

1. How do you know if something has a line of symmetry?

2. What were the shapes that you made with the symmetry sticks?

3. Are these shapes symmetric? How do you know?

4. Where is the line of symmetry in these shapes?

5. What would happen if you taped the shape to the stirrer in a different place?

6. Would the shape you trace be symmetric? How do you know?

Connecting Learning

Part Two

1. How do you find the line of symmetry in a shape?

2. Can a shape have more than one line of symmetry? How do you know?

3. How did you find the lines of symmetry in the shapes you had to cut?

4. How many lines of symmetry did you find in each shape?

5. Did cutting along each of these lines always give you a different result?

6. What did you learn about symmetry from this activity?

Slip Slidin'

Topic
Slides (translations)

Key Questions
1. What is a slide (translation) in geometry?
2. How can a series of slides (translations) be used to move a game piece from the start to the finish on a game board?

Learning Goals
Students will:
- investigate, recognize, and predict the results of motion geometry; and
- apply slides (translations) to a penguin game piece in order to win a game.

Guiding Documents
Project 2061 Benchmark
- *Some features of things may stay the same even when other features change. Some patterns look the same when they are shifted over, or turned, or reflected, or seen from different directions.*

NCTM Standards 2000*
- *Predict and describe the results of sliding, flipping, and turning two-dimensional shapes*
- *Investigate, describe, and reason about the results of subdividing, combining, and transforming shapes*

Math
Geometry
   transformations
      slides

Integrated Processes
Observing
Comparing and contrasting
Applying

Materials
For each student:
   penguin game piece
   penny
   tape
   scissors
   sliding pond

For each pair of students:
   game board
   die

For the class:
   overhead transparency (see *Management 2*)
   overhead game pieces in two colors

Background Information
In geometry, when an object changes position or location, it is called a transformation. The three main ways to change the position or location of an object are by sliding, flipping, or turning it. For this lesson, only slides (translations) will be explored. In a slide (translation), every part of the shape or object slides the same distance in the same direction. The orientation of the shape or object does not change, but its location does.

It is important for students to see how math concepts are not isolated skills but are often used in our everyday lives. Geometry transformation concepts are easily applied to our everyday lives. For example, in real life we often change the location of various objects such as furniture, books, etc., by sliding, flipping, or turning them to create a more pleasing arrangement.

Management
1. Prior to teaching this lesson, copy the sliding ponds and game boards on card stock and laminate for extended use—one game board for each pair of students, one sliding pond per student.
2. Make an overhead transparency of the game board for demonstration purposes.

Procedure
Part One
1. Ask the students what it means for an object to slide (translate) on a surface like the board, a table, or on the floor. Invite a student to demonstrate by sliding an object across the table. Discuss all of the possible ways the object could slide (translate).
2. Explain to students that in geometry, changing an object's position or location is called a transformation.
3. Tell the class that they are going to explore one type of transformation called a slide (translation).

SHAPES, SOLIDS, AND MORE   © 2009 AIMS Education Foundation

4. Give each student a penny, sliding pond page, and penguin. Instruct them to cut out the bold solid rectangle around the penguin and to fold on the dashed line. Have them tape the penny in the appropriate place so that the penguin will stand.
5. Tell the students to place their penguins on the dot in the center of the pond. Draw their attention to the compass rose in the top left corner of the page. Encourage them to use the compass rose if they get confused about which direction their penguins should slide.
6. Ask the students to slide their penguins south. Have them describe the new location. Have students return their penguins to the center. Ask them what might happen if they were to slide their penguins to the east end of the pond. [They could fall through the thin ice.]
7. Continue having students slide their penguins north, south, east, west, northwest, northeast, southeast, and southwest and describe the new locations. On occasion, rephrase the questions such as asking which direction the penguins would have to slide if they wanted to warm by the fire.
8. Have students put their names on their penguins and collect the ponds and penguins.
9. Revisit the terms introduced—transformations and slides (translations). Have students explain each term in their own words.

Part Two
1. Review what a slide (translation) is and the basic directions an object can move.
2. Explain to the class that you will be showing them how to play a game in which they will be changing the location of their penguins by sliding them around the board. Explain that the object of the game is to be the first penguin to reach the finish.
3. Display the game board transparency on the overhead. Place the two colored game pieces on start. Choose one game piece for yourself; the other game piece is for the class. Roll the die. Move one of the markers the number of spaces indicated on the die. (Any combination of horizontal, vertical, and diagonal moves may be used.) Allow a student to roll the die for the class. Tell the class that they may choose which direction they would like to move their marker. Explain that they will only roll the die to move from the start position. On every other turn, they will follow the directions on the space on which they land. Tell the students that if they land on an empty space, they must return to the start and roll the die for their next turn. Play ends when time runs out or a student reaches the finish. If time runs out, the winner is the person closest to the finish.
4. Pair up the students and distribute the game boards, dice, and penguins.
5. Allow time for students to play the game.
6. End with a discussion about when and where in the real world they have transformed an object by sliding it. [rearranging the classroom, moving things on their desks, passing food at the dinner table, etc.]

Connecting Learning
1. What is a slide (translation) in geometry? What directions do objects usually slide (translate)? [north, south, east, west, northwest, southwest, northeast, southeast]
2. When your penguin was in the center of the pond, which direction did it slide (translate) to get refreshments? [NE]
3. If your penguin slid east to the edge of the pond, what might happen? [fall through the thin ice]
4. How did you decide which direction you should go on the game board?
5. When in real life have you changed the location of an object by sliding (translating) it? Describe it.

* Reprinted with permission from *Principles and Standards for School Mathematics*, 2000 by the National Council of Teachers of Mathematics. All rights reserved.

Slip Slidin'

Key Questions

1. What is a slide (translation) in geometry?
2. How can a series of slides (translations) be used to move a game piece from the start to the finish on a game board?

Learning Goals

Students will:

- investigate, recognize, and predict the results of motion geometry; and
- apply slides (translations) to a penguin game piece in order to win a game.

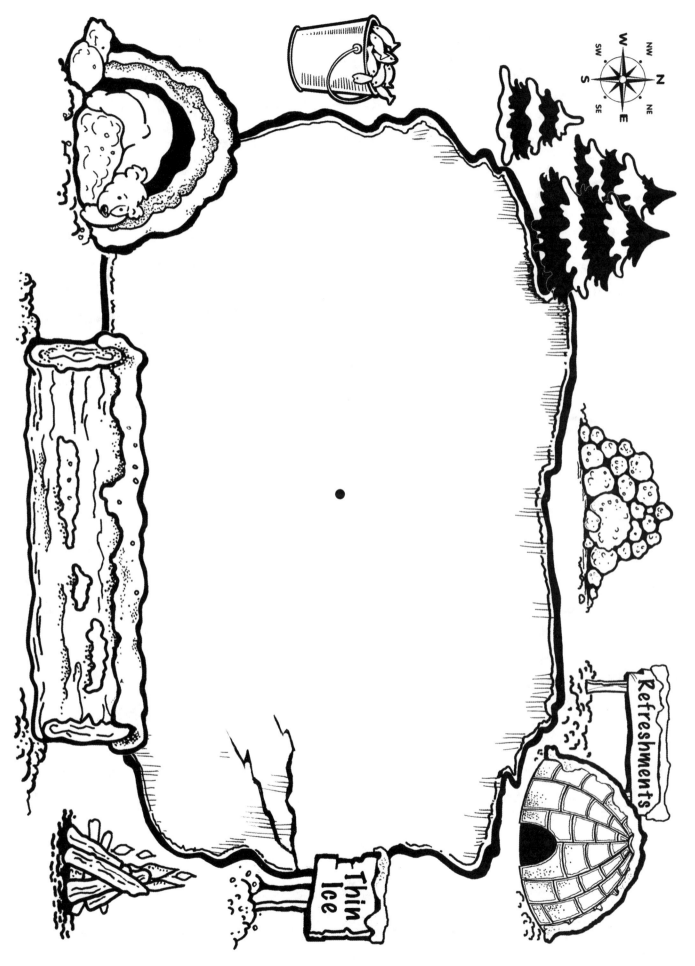

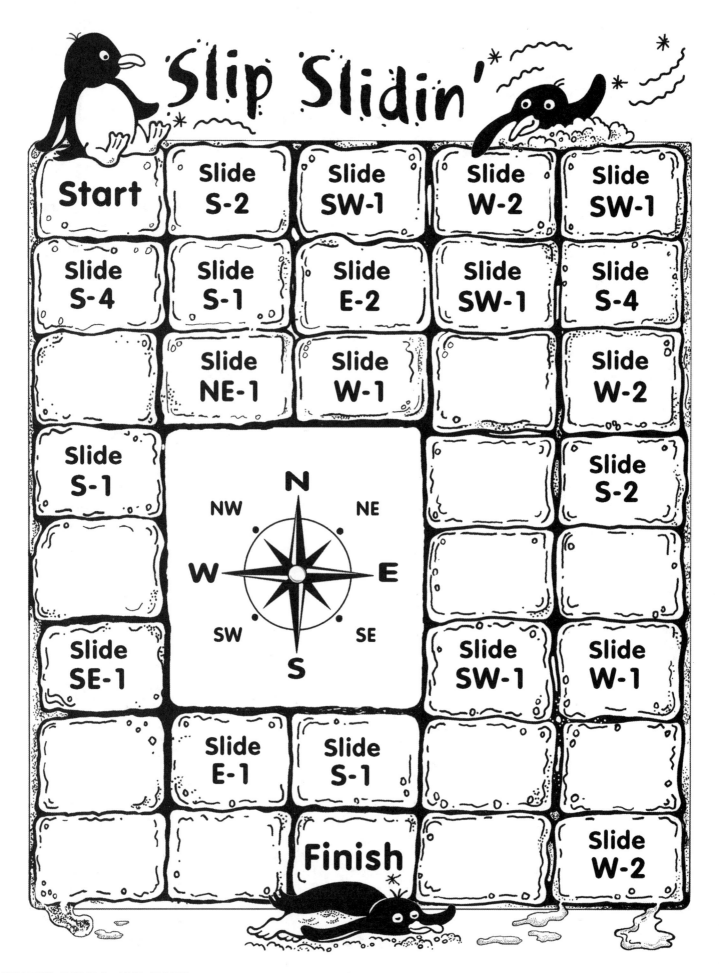

Slip Slidin'

Connecting Learning

1. What is a slide (translation) in geometry? What directions do objects usually slide (translate)?

2. When your penguin was in the center of the pond, which direction did it slide (translate) to get refreshments?

3. If your penguin slid (translated) east to the edge of the pond, what might happen?

4. How did you decide which direction you should go on the game board?

5. When in real life have you changed the location of an object by sliding (translating) it? Describe it.

Flippin' Frogs

Topic
Flips (reflections) and turns (rotations)

Key Question
What changes when we flip (reflect) and turn (rotate) objects?

Learning Goals
Students will:
- use frog cards to demonstrate flips (reflections) and turns (rotations),
- identify and describe the effects after applying the transformations,
- use flips (reflections) and turns (rotations) to complete patterns, and
- identify places where flips (reflections) and turns (rotations) are used in the real world.

Guiding Documents
Project 2061 Benchmark
- *Some features of things may stay the same even when other features change. Some patterns look the same when they are shifted over, or turned, or reflected, or seen from different directions.*

NCTM Standards 2000*
- *Predict and describe the results of sliding, flipping, and turning two-dimensional shapes*
- *Investigate, describe, and reason about the results of subdividing combining, and transforming shapes*

Math
Geometry
  transformations
    flips, turns

Integrated Processes
Observing
Comparing and contrasting
Applying
Predicting

Materials
Flippin' Frogs, one per student
Triangles (see *Management 2*)
Student page
Glue sticks
Scissors

Background Information
Transformations of slides (translations), flips (reflections), and turns (rotations) involve changes in position and/or location. This activity involves flips (reflections) and turns (rotations).

Putting together a jigsaw puzzle is a real-world example of motion geometry that may help students better understand *transformations*. For example, when we get a jigsaw puzzle, we dump all the pieces out of the box onto a table. Then we usually *flip* the pieces over so they are all face up. In the process of completing the puzzle, we often find ourselves *turning* the pieces to the right, then the left, a quarter turn, then a half turn, etc. We may even *slide* a piece or two that we have already assembled across the table to attach to the bigger picture. It can be helpful if students understand that each of these movements represents a *transformation* of the piece, and that each of these has a special name. *Flipping* the piece over is an example of a *reflection* (or *flip*). Changing the angle of the piece is an example of a *rotation* (or *turn*). Moving the piece across the table is an example of a *translation* (or *slide*).

Management
1. When introducing turns (rotations), it may be necessary to explain quarter turns (rotations) (90-degree turn to the left or right) to students. The hands on a clock are a good way to show a quarter turn. It is easiest for students to relate the turns when they begin at the 12 o'clock position.
2. It is suggested that you precut the triangles out of colored paper for each student prior to teaching this lesson. However, you may choose to let students cut them themselves. There are three triangles per page.

Procedure
Part One
1. Share the jigsaw puzzle illustration with the class (see *Background Information*).
2. Give each student a Flippin' Frog. Demonstrate how to fold the page in half so that they have a frog with both a front and back. Instruct students to glue the two halves together.
3. Ask the students to place the frogs on the table in front of them. Tell them to position the frogs so the legs are nearest them and the heads are toward the opposite side of the table. Explain that

they are going to use the frogs to explore two geometry terms—*flipping (reflecting)* and *turning (rotating)*.

4. Have students predict what they would see if the frog card were flipped (reflected) up. Discuss predictions, then ask the students to flip (reflect) the frogs. Repeat the process, this time having the students predict what they would see if the frog were flipped (reflected) down, to the left, and to the right.
5. When students are comfortable with flips (reflections), give each a construction paper triangle. Discuss how flipping (reflecting) a triangle would be different than flipping (reflecting) a frog.
6. Play "Simon Says" for students to explore flipping (reflecting) the triangles. For example, "Simon says, flip (reflect) your triangle up. Flip (reflect) your triangle down."
7. Instruct the students to place the frogs and triangles in a safe place for use in *Part Two* of the activity.
8. End with a discussion about how flipping (reflecting) changes an object's location and direction and how flipping (reflecting) shapes is or isn't different than flipping (reflecting) frogs.

Part Two
1. Review what it means to flip (reflect) an object. Revisit the jigsaw illustration as it relates to turns (rotations).
2. Have students again place their frogs on the table in front of them. Tell them to position the frogs so the legs are nearest them and the heads are toward the opposite side of the table. Ask students to predict what the frog would look like if they turned (rotated) it a quarter turn to the right. (It may be necessary to use the clock to illustrate quarter turns.) Tell the students to turn (rotate) the frogs to check their predictions.
3. Repeat procedure two turning (rotating) left, right, a quarter turn, a half turn, etc.
4. Discuss how turning (rotating) would be different if they were turning shapes instead of frogs. Revisit turns (rotations) using the triangles from *Part One* of this lesson.
5. Play "Simon Says" using the triangles as a review of turns (rotations).

Part Three
1. Ask students to describe how flips (reflections) and turns (rotations) change the locations and directions of objects and shapes.
2. On the board, draw the triangle pattern shown below.

3. Ask students what is missing in the pattern and have them identify the transformations as flips (reflections) or turns (rotations). Most students will probably see this as a flip pattern; however, it could also be described as a turn pattern, with the triangle turning (rotating) a half turn each time. Both would be acceptable answers as long as the students can justify their answers.
4. Place the following pattern on the board for the students to describe.

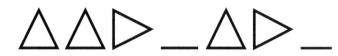

5. Challenge them to identify what is missing and describe the transformations used. In this case, the pattern is flip (reflect), turn (rotate), turn (rotate), flip (reflect), turn (rotate), turn (rotate)...
6. Give each student the pattern pages and challenge them to fill in the missing parts to the patterns and to describe the transformations that are taking place in the patterns.
7. End with a discussion about patterns made using flips (reflections) and turns (rotations) of a shape or object such as quilts, tile, wallpaper etc.

Connecting Learning
1. What changed when you flipped (relected) the frog?
2. How was it different when we flipped (reflected) triangles instead of frogs?
3. What changes when we turn (rotate) an object?
4. Where is the real world do we see the use of flips (reflections) and turns (rotations)?

* Reprinted with permission from *Principles and Standards for School Mathematics*, 2000 by the National Council of Teachers of Mathematics. All rights reserved.

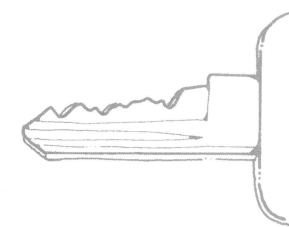

Key Question

What changes when we flip (reflect) or turn (rotate) an object?

Learning Goals

Students will:

- use frog cards to demonstrate flips (reflections) and turns (rotations),
- identify and describe the effects after applying the transformations,
- use flips (reflections) and turns (rotations) to complete patterns, and
- identify places where flips (reflections) and turns (rotations) are used in the real world.

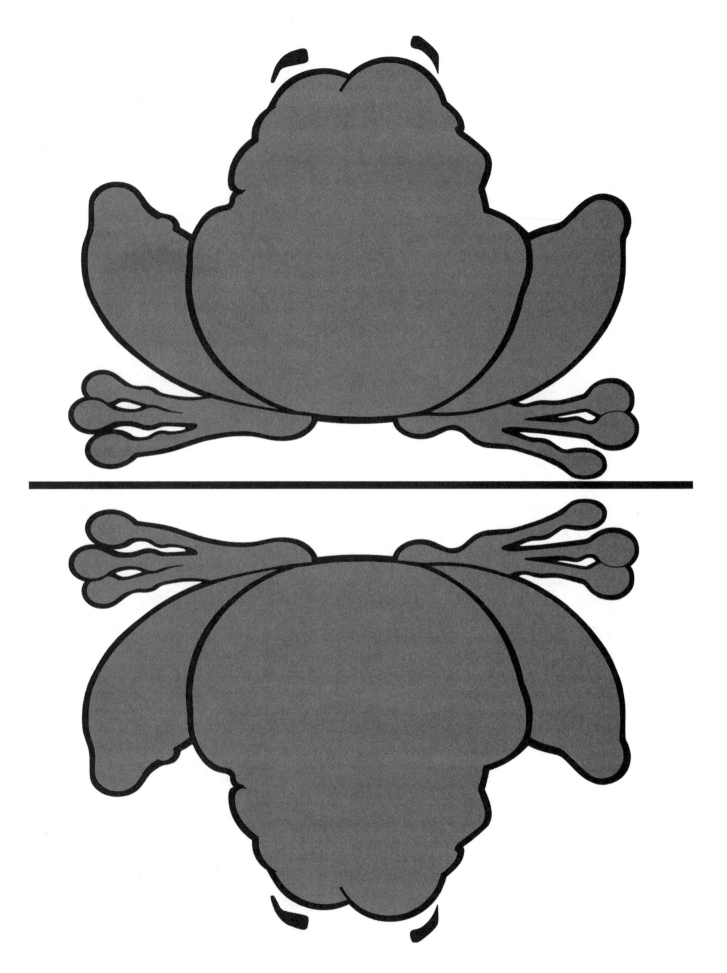

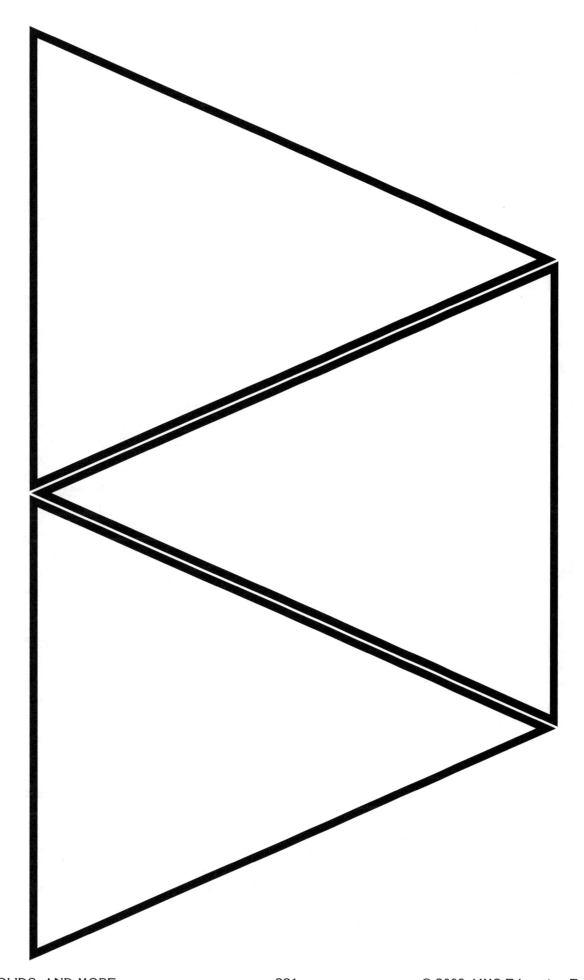

Flippin' Frogs

Fill in the missing parts to these patterns.

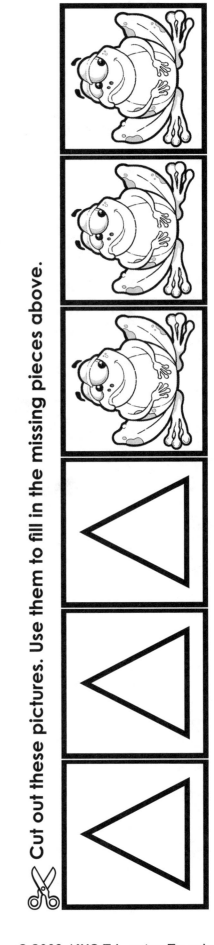

Cut out these pictures. Use them to fill in the missing pieces above.

Flippin' Frogs

Connecting Learning

1. What changed when you flipped (relected) the frog?

2. How was it different when we flipped (reflected) triangles instead of frogs?

3. What changes when we turn (rotate) an object?

4. Where is the real world do we see the use of flips (reflections) and turns (rotations)?

Flipping for Transformations

Topic
Flips (reflections)

Key Question
What does it look like when a triangle is flipped (reflected)?

Learning Goal
Students will construct flip books to show triangles being transformed by flips (reflections).

Guiding Document
*NCTM Standards 2000**
- *Recognize and apply slides, flips, and turns*
- *Predict and describe the results of sliding, flipping, and turning two-dimensional shapes*
- *Investigate, describe, and reason about the results of subdividing, combining, and transforming shapes*

Math
Geometry
  transformations
    flips

Integrated Processes
Observing
Comparing and contrasting
Applying

Materials
Part One:
  pattern blocks
  colored yarn (see *Management 1*)
  12" x 18" construction paper (see *Management 2*)
  glue
  colored pencils

Part Two:
  triangles (see *Management 3*)
  flip book pages (see *Management 4*)
  scissors
  rubber cement
  stapler
  masking tape

Background Information
Flips, also called reflections, are transformations in which a shape or other object is reflected across a line. This line can be anywhere in relation to the shape, including within the shape. In this activity, students will create a flip book to illustrate the motion of an equilateral triangle flipping (reflecting) back and forth across a line. In this case, the line is adjacent to one side of the triangle.

Management
1. Use a thin yarn that is a different color than the construction paper. Cut the yarn into 10-inch lengths. Each student needs three pieces of yarn.
2. Select white or light-colored construction paper so that students will be able to see the shape outlines that they trace on the paper.
3. Each student will need 30 paper equilateral triangles with a side length of one inch. (These are the same size as the equilateral triangle shape in a set of pattern blocks.) These triangles can be cut from colored paper using a die-cut machine, or the provided page can be copied on colored paper and students can cut out the required number of triangles. Be sure to cut the triangles from colored paper so that they will be visible in the flip book.
4. Each student needs two copies of the same flip book page. There are two different pages provided so that some students will have a side-to-side flip and others will have an up-and-down flip. You may choose to use a paper cutter to cut these pages apart before the activity, or students may do the cutting themselves. The paper needs to be cut all the way to its edges along the solid lines to make 15 small rectangular pages that are all the same size and shape.

Procedure
Part One
1. Ask the students what a flip (reflection) is. Invite several students to show you examples of flips (reflections) using objects in and around their desks. Tell students that a flip is a reflection across a line, but the line doesn't have to be touching the shape or object.
2. Give students a sheet of construction paper, labels, scissors, glue, and three pieces of yarn. Have them cut apart the labels.

SHAPES, SOLIDS, AND MORE © 2009 AIMS Education Foundation

3. Instruct students to orient their construction paper so that the long edge is facing them and glue the labels to the paper so that they are evenly spaced along the top of the paper (the long edge). Have them glue the pieces of yarn to the paper so that they form three straight lines that are under the three labels.

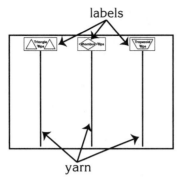

4. Have students set aside their papers to let the glue dry. Explain that once their papers have dried, their challenge will be to find and record as many different flips (reflections) using the triangle, rhombus, and trapezoid pattern block shapes as they can. They will record their solutions by tracing the pattern block and its flip (reflection) on either side of the pieces of yarn, which are the flip (reflection) lines.

5. Discuss the ways that the shapes can be placed in relation to the lines. The shapes must touch the line either at a corner or along a side. No part of the shape can cross the line. If the shape touches at a corner, it must be oriented so that it is either parallel or perpendicular to the line, no strange angles. Use a triangle pattern block on the overhead to show students the following examples of the two correct corner orientations and others that are incorrect. Discuss why there might be rules to how the shapes can touch the line. [If there were no restrictions, the number of possible flips (reflections) would be infinite.]

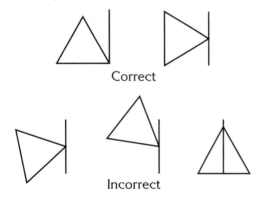

6. When the glue is dry, distribute the paper, the colored pencils, and the pattern blocks. Allow time for students to explore with their pattern blocks to find as many flips (reflections) as they can with each shape. Have them trace each flip (reflection) on the line under the correct label.

7. Once students are done, draw three vertical lines on the board. Label the lines *Triangle Flips (Reflections)*, *Rhombus Flips (Reflections)*, and *Trapezoid Flips (Reflections)*. Invite students to come to the front and trace one solution they discovered on the board. Continue until all of the solutions students discovered have been recorded.

8. Determine if all of the solutions have been discovered, and add any that may be missing. Also discuss any solutions that may be duplicates of each other, just seen from a different perspective. [There are 11 solutions total.]

Part Two

1. Distribute two flip book pages to each student. Explain that they will be making flip books to show one way that a triangle can be transformed by flipping (reflecting) it across a line.

2. If the flip book pages are not already cut apart, have students cut along the solid lines to make 30 small rectangular pages. Instruct them to stack the pages so that they are all oriented the same way—with the horizontal (or vertical) line closest to the right edge of the paper.

3. Distribute the triangles (or page of triangles) to each student. If necessary, have the students cut apart the triangles. Have them count to verify that they have 30 triangles.

4. Tell students that they will be gluing the triangles to the pages to make a flip book. In this case, the triangle must touch the line along a side, not just at a corner. Each page of the book must show the flip of the triangle from the previous page. Discuss what this would look like with each line orientation (horizontal and vertical).

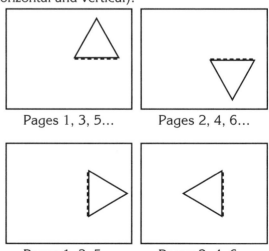

5. Distribute the rubber cement and have students carefully glue down the triangles, one page at a time. As they glue each page, instruct them to set the pages aside to dry so that they are not stacked on top of each other.
6. After the triangles are all glued down and the glue is dry, have students rub off any glue residue that may be around the edges of the triangles. This will prevent the pages of the book from sticking together when the flip book is assembled.
7. Instruct students to stack the pages of their flip books in order so that each page shows the flip of the page before. When students get their books in order, have them bring the pages to you.
8. Staple each student's book along the left edge, being sure that the pages are flush along the right side. Put a piece of masking tape over the spine of the book to cover the staples.

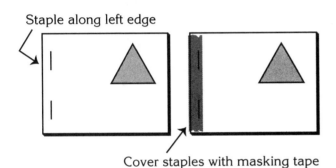

9. Show students how to hold the spine of the book in on one hand and use the thumb of the opposite hand to flip quickly through the pages. If assembled correctly, it should look like the triangle is flipping side to side or up and down. Encourage students to trade books with others who have a different orientation to see both variations.

Connecting Learning
1. What is a flip (reflection)?
2. How many different ways were you able to flip (reflect) the triangle? ...the rhombus? ...the trapezoid?
3. Do you think you have found them all? Why or why not?
4. How did you know how to glue down the triangles for the pages of your flip book?
5. What do you see when you flip the pages of the flip book?
6. What have you learned about flips (reflections) from this activity?

Solutions
Given the restrictions specified (shapes must touch along an edge or at a corner, shapes must not overlap the line, shapes must be parallel or perpendicular to the line) there are only 11 possible solutions using the three shapes provided.

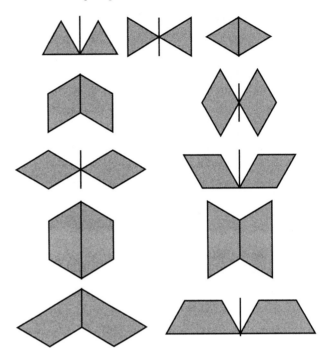

* Reprinted with permission from *Principles and Standards for School Mathematics*, 2000 by the National Council of Teachers of Mathematics. All rights reserved.

Flipping for Transformations

Key Question

What does it look like when a triangle is flipped (reflected)?

Learning Goal

Students will:

construct flip books to show triangles being transformed by flips (reflections).

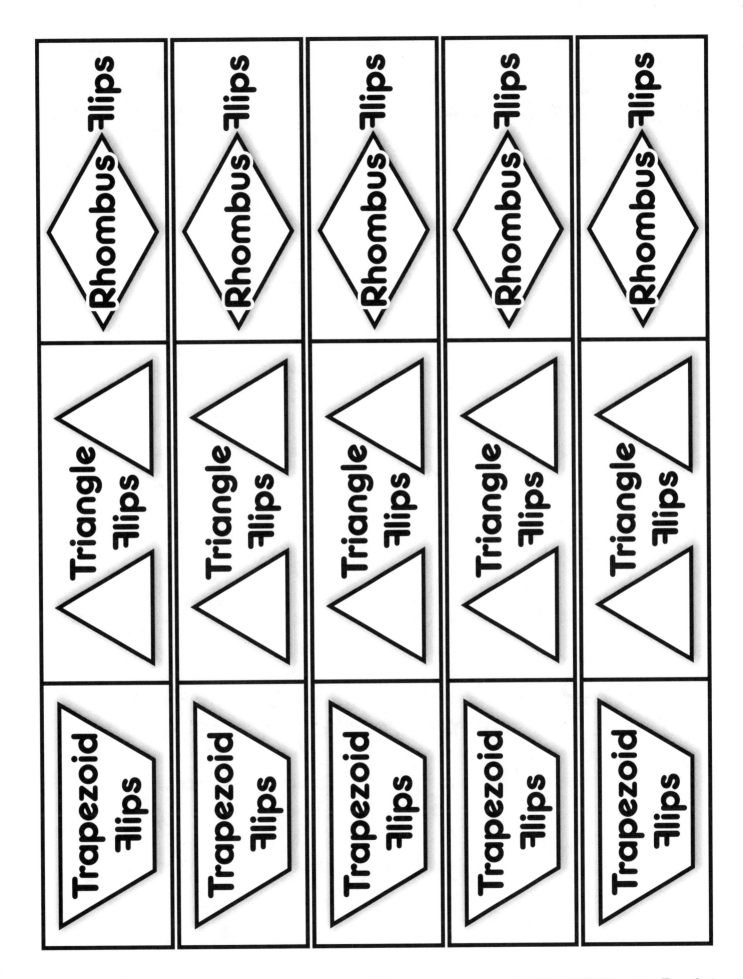

SHAPES, SOLIDS, AND MORE 229 © 2009 AIMS Education Foundation

SHAPES, SOLIDS, AND MORE 230 © 2009 AIMS Education Foundation

SHAPES, SOLIDS, AND MORE

GEOMETERY 232 © 2009 AIMS Education Foundation

Flipping for Transformations

Connecting Learning

1. What is a flip (reflection)?

2. How many different ways were you able to flip (reflect) the triangle? …the rhombus? …the trapezoid?

3. Do you think you have found them all? Why or why not?

4. How did you know how to glue down the triangles for the pages of your flip book?

5. What do you see when you flip the pages of the flip book?

6. What have you learned about flips (reflections) from this activity?

Picturing Rotations

Topic
Turns (rotations)

Key Question
What are some different ways that shapes can be turned or rotated?

Learning Goals
Students will:
- recognize that a rotation is a turn around a point, and
- make rotation drawings by tracing pattern blocks that have been turned around a point.

Guiding Document
*NCTM Standards 2000**
- *Recognize and apply slides, flips, and turns*
- *Predict and describe the results of sliding, flipping, and turning two-dimensional shapes*
- *Investigate, describe, and reason about the results of subdividing, combining, and transforming shapes*

Math
Geometry
    transformations
        turns

Integrated Processes
Observing
Comparing and contrasting
Relating

Materials
For each student:
    coffee stirrer (see *Management 1*)
    sticky tacky (see *Management 2*)
    pattern block
    blank paper
    colored pencil

Background Information
    Turns are transformations in which a shape or object is turned, or rotated, around a point (or line). This point (or line) can be anywhere in relation to the shape or object, including within the object itself. Students in primary grades are most commonly exposed to examples of geometric shapes rotating around the center point, but they should recognize that turns around corners, edges, or even points outside the shape are also rotations. This activity will give them the opportunity to create visual images of rotations by turning shapes around different points and tracing their outlines.

Management
1. Be sure you get plastic coffee stirrers that are round, like miniature straws.
2. Clay can be substituted for sticky tacky as long as it will hold the coffee stirrer firmly in place on the pattern block.

Procedure
1. Invite a student to the front of the room. Ask him or her to turn around. Ask him or her to turn to the right, then to the left, halfway, all the way, etc.
2. Ask the class to describe what the student just did. How did he/she know what to do when he/she was asked to turn around? What does it mean to turn?
3. Invite a second student to the front of the room. Put a chair in the middle of an open area and ask that student to place a hand on the back of the chair and then turn around the chair without removing his/her hand. Repeat the process from the first example by having the student turn right, left, halfway, etc.
4. Have the class compare the two turns. What things were the same? What things were different? Discuss how in both cases, the student was turning around a point. In one case, that point was his/her hand on the back of the chair. In the other case, the point was not clearly visible, being somewhere in the center of the student's body as he/she turned in one place.
5. Explain that students are going to explore what happens when shapes are turned or rotated in different ways. Give each student a coffee stirrer and a small lump of sticky tacky. Have them select a pattern block shape.

SHAPES, SOLIDS, AND MORE   235   © 2009 AIMS Education Foundation

6. Instruct students to place the lump of sticky tacky somewhere on the shape and then stick the coffee stirrer into the lump of sticky tacky.
7. Invite them to twirl the coffee stirrer between their thumbs and fingers and to observe what happens to the shape. [It turns (rotates).] Have them spin the shape slow, fast, to the right, to the left, etc.
8. Tell students that they are now going to have the chance to make pictures by turning (rotating) their shapes around different points. Give students a sheet of plain white paper and a colored pencil.
9. Instruct the students to place their pattern block shapes on the paper and carefully trace around them.
10. Have the students lift their shapes from the paper and mark the locations of the coffee stirrers. Explain that this is marking the point around which the shape will be turning (rotating).
11. Using a shape on the overhead, show students how to turn the shape by turning the coffee stirrer so that the stirrer stays in the same place, but the shape turns. Rotate it less than 90°, trace around the shape, rotate it again, trace it again, and rotate it one more time and trace so that there is a total of four tracings of the shape. Explain that this is what students are to do on their papers.
12. Allow time for students to make multiple drawings of their shapes on the paper, each time putting the coffee stirrer in a different location on the pattern block so that they can compare how the different rotation pictures look.

Connecting Learning
1. What does it mean to turn (rotate)? How did _____ (name of student) demonstrate a turn (rotation)?
2. How was the way _____ (name of student) turned similar to the way that _____ (name of student) turned? How was it different? [One turn was small, the person stayed almost in the same place, etc. The other turn was bigger, the person made a large circle, etc.]
3. What ways did you pick to turn (rotate) your pattern block shape?
4. How do your different pictures compare? What things are the same? What things are different?
5. Which of your pictures is most like when _____ (name of student) turned in one spot? Why? [Any picture where the shape was rotated from a point inside the shape (as opposed to a point along an edge or at a corner) is like the student who turned in place.]
6. Which of your rotation pictures is most like when _____ (name of student) turned around the chair? Why? [Any picture where the shape was rotated from a corner is like the student who turned around the chair.]

7. Which picture is your favorite? Why?
8. What things in the real world turn (rotate)? [tires, desk chairs, merry-go-round, ferris wheel, a globe, etc.]

Solutions
Some of the rotation pictures your students might create are pictured here. There are many others.

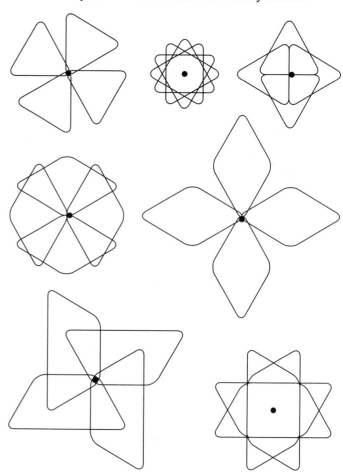

* Reprinted with permission from *Principles and Standards for School Mathematics*, 2000 by the National Council of Teachers of Mathematics. All rights reserved.

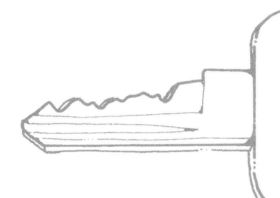

Key Question

What are some different ways that shapes can be turned or rotated?

Learning Goals

Students will:

- recognize that a rotation is a turn around a point, and
- make rotation drawings by tracing pattern blocks that have been turned around a point.

Connecting Learning

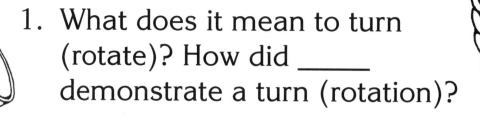

1. What does it mean to turn (rotate)? How did _____ demonstrate a turn (rotation)?

2. How was the way _____ turned similar to the way that _____ turned? How was it different?

3. What ways did you pick to turn (rotate) your pattern block shape?

4. How do your different pictures compare? What things are the same? What things are different?

5. Which of your pictures is most like when _____ turned in one spot? Why?

Picturing Rotations

Connecting Learning

6. Which of your rotation pictures is most like when _____ turned around the chair? Why?

7. Which picture is your favorite? Why?

8. What things in the real world turn (rotate)?

The Pattern Block Shuffle

Topic
Transformations

Key Question
What is the difference between sliding (translating), flipping (reflecting), and turning (rotating) a pattern block?

Learning Goals
Students will:
• slide (translate), flip (reflect), and turn (rotate) various pattern blocks; and
• identify and describe the geometric shapes after applying the transformations.

Guiding Documents
Project 2061 Benchmark
• *Many objects can be described in terms of simple plane figures and solids. Shapes can be compared in terms of concepts such as parallel and perpendicular, congruence and similarity, and symmetry. Symmetry can be found by reflection, turns, or slides.*

NCTM Standards 2000*
• *Recognize and apply slides, flips, and turns*
• *Predict and describe the results of sliding, flipping, and turning two-dimensional shapes*

Math
Geometry
 transformations
  slides, flips, turns

Integrated Processes
Observing
Communicating
Collecting and recording data
Applying

Materials
Colored card stock (see *Management 1*)
Pattern blocks (see *Management 2*)
Tape or sticky tacky
Student page, optional

Background Information
Often students are asked to identify how a shape has been transformed (flipped, slid, or turned) to achieve its current position. To determine this, they have to mentally manipulate the shape on the page. In order to develop the ability to mentally manipulate an object, students first need experiences in which they explore the transformations by physically flipping (reflecting), sliding (translating), and turning (rotating) shapes.

Transformations change the position of a shape without changing its form or size. The following terms are used to describe how a shape has been transformed.

Flips (reflections)—When a figure is flipped, it is reflected across a line. It can be flipped left or right or up and down. The flipped figure will be a mirror reflection of the original. Its size and shape do not change.

Slides (translations)—When you slide a figure, it is moved in a straight line to a different place. Its shape, orientation, and size do not change, only its location.

Turns (rotations)—When a figure is turned, it is rotated around a point. The size and shape do not change, but the orientation does.

Management
1. Prior to teaching this lesson, copy the included set of large shapes onto card stock and laminate for extended use. If available, copy the shapes on card stock that corresponds to the colors of the pattern block pieces. Cut out the shapes. The number of copies of each shape should be determined by the number of students you have and the amount of open space at the front of the classroom for the shuffle dance.
2. You will need only the orange square, red trapezoid, green triangle, and blue rhombus pattern block shapes. Each student needs one of each shape.
3. Tape, magnets, or sticky tacky can be used to adhere the shapes to the board in *Part One*.
4. Demonstrate how students should hold the shape cards from the bottom corners when doing the *Pattern Block Shuffle* in *Part Two* of this activity.
5. Prior to doing the *Pattern Block Shuffle* in *Part Two*, inform the students which direction they should slide.

SHAPES, SOLIDS, AND MORE © 2009 AIMS Education Foundation

6. *Part Three* is an optional student page and can be used as an assessment. For students who struggle with identifying the transformation, you may choose to allow them to use the pattern blocks to act out the movements.

Procedure
Part One
1. Ask the students if they have ever gone down a slide, seen someone do a back flip, or turned around in line. Invite students to demonstrate each, and explain that in geometry, slides (translations), flips (reflections), and turns (rotations) are called *transformations*.
2. Using tape or sticky tacky, place a large equilateral triangle on the board so that it is pointing up. Trace around it so that students can be reminded of its original position. Draw a line above the shape and invite a student to flip the card up over the line and to stick it to the board in the flipped position. Discuss how the position was changed but the shape's size and shape have remained the same.
3. Place the triangle back in its original position and invite a student to flip (reflect) it down and then toward one side. Discuss how the flipped (reflected) positions are different from the original.
4. Ask students what would be different if they used a different shape such as a trapezoid instead of a triangle.
5. Explore slides (translations) and turns (rotations) using the same processes used for flips.

Part Two
1. Review the slides, flips, and turns introduced in *Part One* of this activity.
2. Explain to students that you are going to teach them a new dance called the *Pattern Block Shuffle*. Tell them that they will get to practice sliding, flipping, and turning shapes in this dance.
3. Give several students large triangle cards and ask them to stand at the front of the class facing their classmates.
4. Ask them to flip (reflect) their triangles and discuss the different ways that this could be done. Ask them to turn (rotate) their triangles and discuss how they look. Have them slide (translate) their triangles by taking a step to the left (or right).
5. Tell the class that these are the steps in the *Pattern Block Shuffle*. Explain that you will be calling out several steps at once and that they are to follow your verbal directions by acting out the steps.
6. Instruct the students at the front of the class to, "flip it, flip it, slide, slide, flip it, flip it, slide, turn it, turn it, turn it, stop." (Use vocabulary of translate, rotate, and reflect if appropriate.)

7. Give those students who are seated a set of pattern blocks (see *Management 2*). Try it a second time, this time allowing the students at their seats to do the same motions with the corresponding pattern block.
8. Repeat the *Pattern Block Shuffle* using different shape cards and students each time. After each dance, discuss the positions of the shapes and allow those at their seats to compare their results.

Part Three
1. Distribute the student page and ask students to identify the transformations.
2. When all students have completed the student page, discuss their answers. Have students compare the experience of using the pattern block shapes to answering the questions on the page without having the shapes.

Connecting Learning
1. When you look at a shape on a page, how can you tell whether it has been flipped (reflected)? …slid (translated)? …turned (rotated)?
2. What would an equilateral triangle look like if you flip (reflect) it up?
3. What is the difference between sliding (translating), flipping (reflecting), and turning (rotating) a shape?
4. Why do you think we did the *Pattern Block Shuffle*?
5. Was it difficult or easy to identify the slides (translations), flips (reflections), and turns (rotations) on the student page? Explain your answer.

Solutions
1. c
2. a
3. b
4. a
5. a
6. a

* Reprinted with permission from *Principles and Standards for School Mathematics*, 2000 by the National Council of Teachers of Mathematics. All rights reserved.

The Pattern Block Shuffle

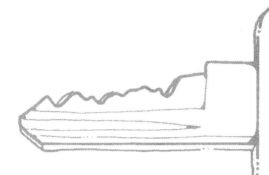

Key Question

What is the difference between sliding (translating), flipping (reflecting), and turning (rotating) a pattern block?

Learning Goals

Students will:

- slide (translate), flip (reflect), and turn (rotate) various pattern blocks; and
- identify and describe the geometric shapes after applying the transformations.

Copy onto orange card stock and cut out.

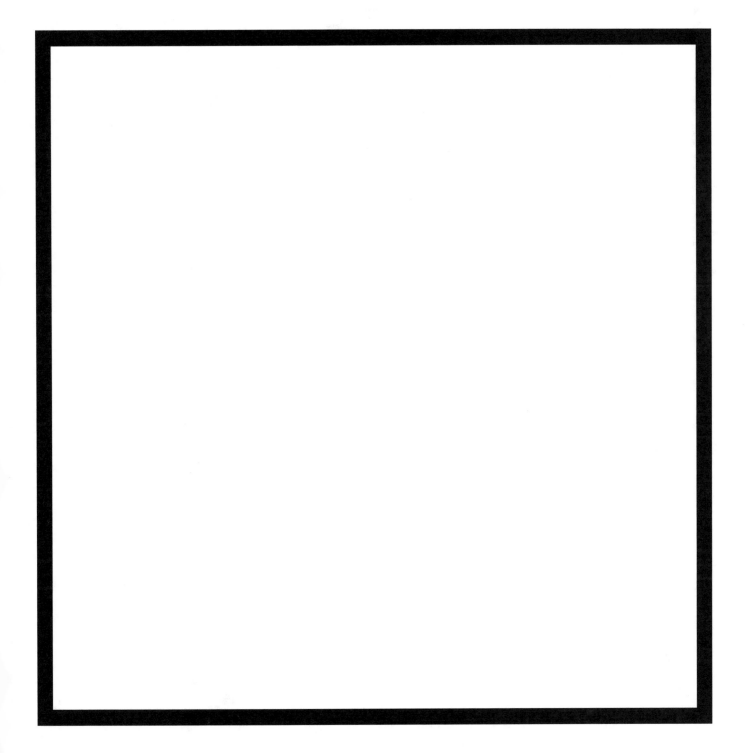

The Pattern Block Shuffle

Copy onto blue card stock and cut out.

SHAPES, SOLIDS, AND MORE 246 © 2009 AIMS Education Foundation

The Pattern Block Shuffle

Copy onto red card stock and cut out.

SHAPES, SOLIDS, AND MORE © 2009 AIMS Education Foundation

Which pair of figures shows a flip over the line?

    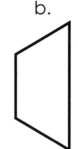

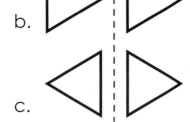    

Which pair of figures shows a slide across the line?

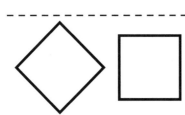    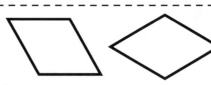

Which pair of figures shows a turn?

Connecting Learning

1. When you look at a shape on a page, how can you tell whether it has been flipped (reflected)? …slid (translated)? …turned (rotated)?

2. What would an equilateral triangle look like if you flip (reflect) it up?

3. What is the difference between sliding (translating), flipping (reflecting), and turning (rotating) a shape?

4. Why do you think we did the *Pattern Block Shuffle*?

5. Was it difficult or easy to identify the slides (translations), flips (reflections), and turns (rotations) on the student page? Explain your answer.

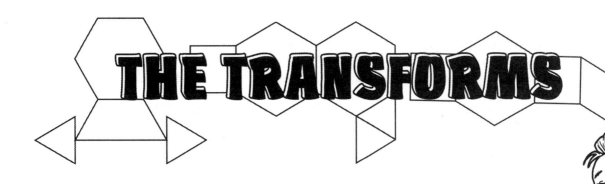

THE TRANSFORMS

Topic
Transformations

Key Question
How can we use transformations to make pictures?

Learning Goal
Students will identify whether pattern block pieces have been slid (translated), flipped (reflected), or turned (rotated) in order to make a picture.

Guiding Documents
Project 2061 Benchmark
- *Some features of things may stay the same even when other features change. Some patterns look the same when they are shifted over, or turned, or reflected, or seen from different directions.*

NCTM Standards 2000
- *Predict and describe the results of sliding, flipping, and turning two-dimensional shapes*
- *Investigate, describe, and reason about the results of subdividing, combining, and transforming shapes*

Math
Geometry
  transformations
    slides, flips, turns

Integrated Processes
Observing
Comparing and contrasting
Drawing conclusions
Applying

Materials
Pattern blocks (see *Management 1*)
Transform mats
Student page
Transparency of Transform Man 1

Background Information
In geometry, transformations describe how a figure or object is moved in a plane. In this activity, three types of transformations are addressed—slides (translations), flips (reflections), and turns (rotations). In a slide (translation), every part of the shape or object slides the same distance in the same direction. The orientation of the shape or object does not change, but its location does. Flips, also called reflections, occur when a shape or object flips over a line, sometimes called the mirror line. Each part of the object or shape is the same distance from the line as its reflection is on the other side. Both its orientation and its location change. Turns, also called rotations, occur when the object or shape is rotated around a point. It can be a point within the object or shape or a point outside of the object or shape. Turns can be clockwise or counterclockwise and are measured in degrees. In a turn, the orientation and sometimes also the location change.

Management
1. To construct all four of the Transforms, each student will need the following pattern blocks: six hexagons, six triangles, three trapezoids, three squares, and one blue rhombus.
2. Prior to teaching this lesson, copy Transform Man 1 onto an overhead transparency for demonstration purposes.

Procedure
1. Review the terms slides (translations), flips (reflections), and turns (rotations).
2. Introduce the four Transforms to the class. Ask the students where they got their names. [Each is created by pattern blocks that have gone through transformations.]
3. Explain to the students that they will be given a series of four pictures (two Transform men and two Transform women), a student recording page, and a set of pattern blocks.

SHAPES, SOLIDS, AND MORE   251   © 2009 AIMS Education Foundation

4. Place the transparency of Transform Man 1 on the overhead. Ask the students what pattern blocks they would need to construct Transform Man 1. [one yellow hexagon, one red trapezoid, two green triangles]
5. To begin, invite a student to place the pattern blocks in the shapes that are outlined with broken lines. Tell that student to transform each shape into the Transform Man. Ask the students to describe the transformation that the trapezoid went through to get below the hexagon. Demonstrate how the student information should be recorded onto the student page.

Connecting Learning
1. What do transformations do to shapes or objects? [change their position and or location]
2. Describe a slide (translation). ...a flip (reflection). ...a turn (rotation).
3. The hexagon on Transform Man 1 can be transformed by sliding (translating). What other ways can it be transformed? [flipped (reflected) or turned (rotated)]
4. What other pieces can be transformed in more than one way?

* Reprinted with permission from *Principles and Standards for School Mathematics*, 2000 by the National Council of Teachers of Mathematics. All rights reserved.

TRANSFORM MAN 1

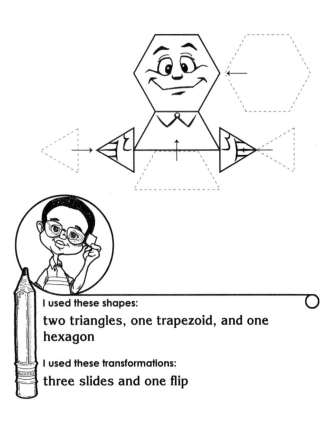

I used these shapes:
two triangles, one trapezoid, and one hexagon

I used these transformations:
three slides and one flip

6. Give each student a set of pattern blocks, a copy of the student page, and each of the four Transform Mats.
7. Allow students time to work through the four transformation mats and to record their movements.
8. End with a discussion about transformation. Challenge students to create their own Transform Man and/or Transform Ma'am.

THE TRANSFORMS

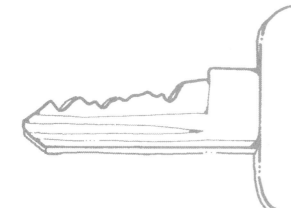

Key Question

How can we use transformations to make pictures?

Learning Goal

Students will:

identify whether pattern block pieces have been slid (translated), flipped (reflected), or turned (rotated) in order to make a picture.

TRANSFORM MAN 1

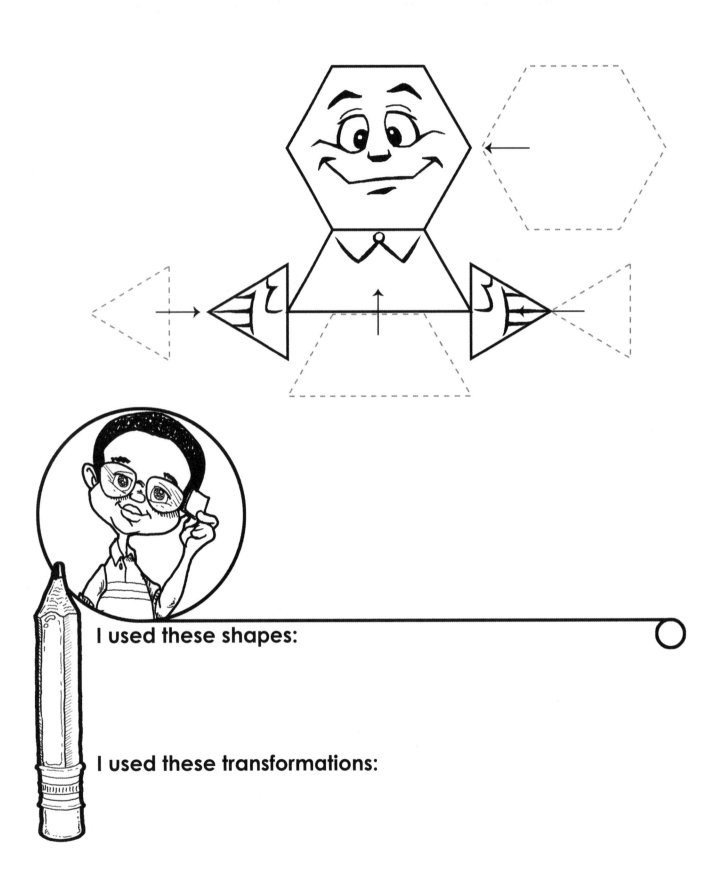

I used these shapes:

I used these transformations:

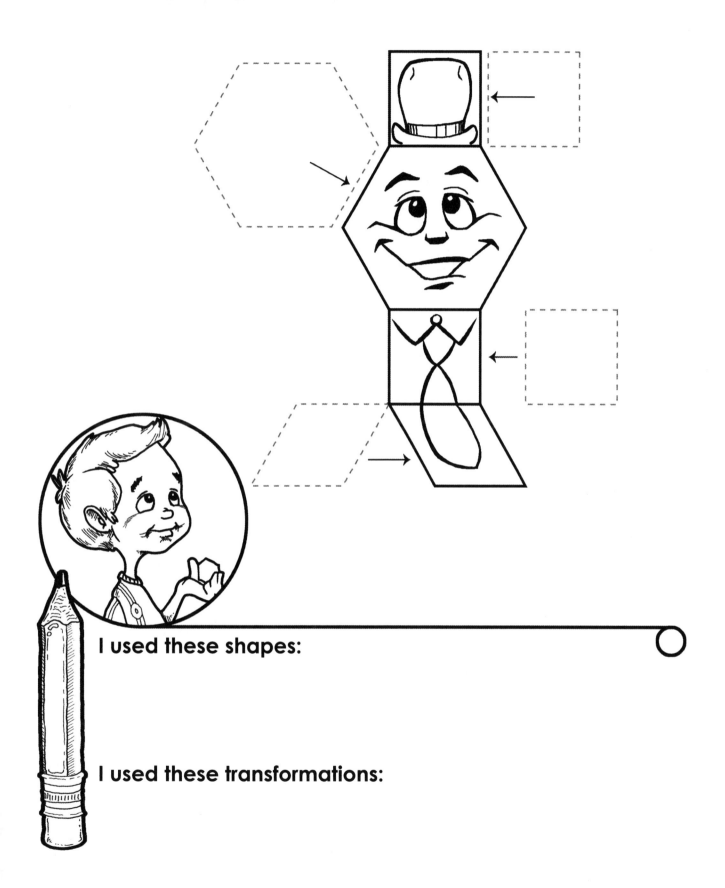

TRANSFORM MA'AM 1

I used these shapes:

I used these transformations:

SHAPES, SOLIDS, AND MORE

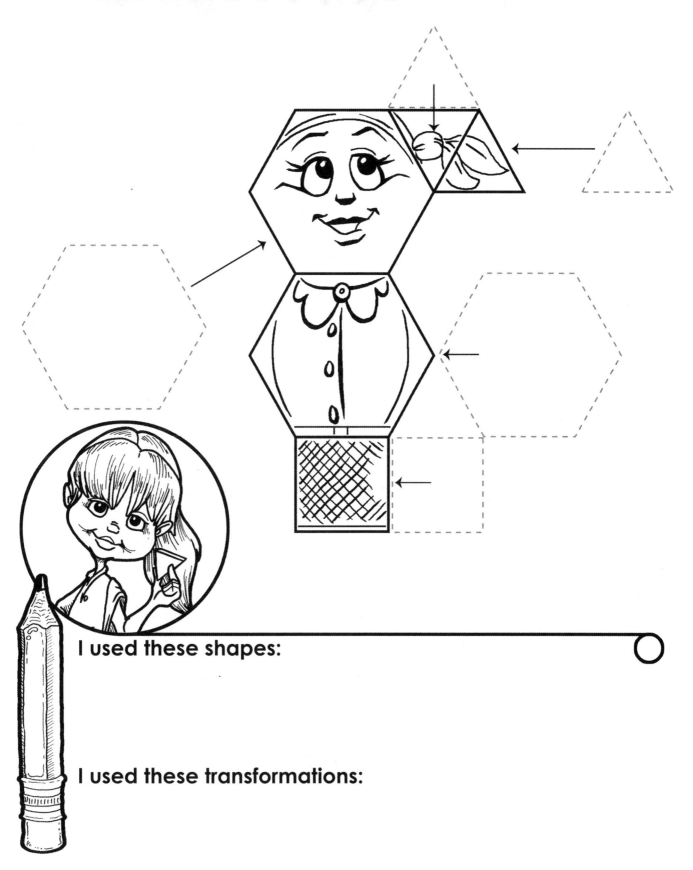

THE TRANSFORMS

Connecting Learning

1. What do transformations do to shapes or objects?

2. Describe a slide (translation). ...a flip (reflection). ...a turn (rotation).

3. The hexagon on Transform Man 1 can be transformed by sliding (translating). What other ways can it be transformed?

4. What other pieces can be transformed in more than one way?

Quilted Transformations

Topic
Transformations

Key Question
How can you use transformations to make quilt squares?

Learning Goal
Students will apply transformations to shapes as directed by the teacher to create quilt squares.

Guiding Document
NCTM Standards 2000*
- *Recognize and apply slides, flips, and turns*
- *Describe location and movement using common language and geometric vocabulary*
- *Predict and describe the results of sliding, flipping, and turning two-dimensional shapes*

Math
Geometry
    transformations
        slides, flips, turns
    2-D shapes

Integrated Processes
Observing
Comparing and contrasting
Applying

Materials
Paper triangles (see *Management 2*)
Quilt square pages
Scissors
Glue sticks

Background Information
    Turns (rotations), flips (reflections), and slides (translations) are the three transformations that can be applied to two-dimensional geometric shapes. This activity will give students the opportunity to apply these transformations in the context of making quilt squares. They will hear geometric vocabulary used in context as they listen and apply directions given by the teacher. The final product of colorful quilt squares can be used to address other geometric topics such as symmetry.

Management
1. This activity is not intended to be an introduction to transformations. It is expected that students have already had multiple experiences where they have learned to identify and perform transformations.
2. Make enough copies of the page of triangles provided so that each student can have 20 triangles. Make the copies on at least two different colors of paper. To save time in class, these triangles can be cut apart prior to doing the activity. Otherwise students can cut the triangles apart in class themselves.
3. The following numbers and colors of triangles are needed:
Turns (rotations) quilt square: four triangles of the same color
Flips (reflections) quilt square: eight triangles of the same color
Slides (translations) quilt square: four triangles of one color and four of a second color
4. Students do not have to make identical color designs; they can choose their colors for the different quilt squares.

Procedure
1. If students will be cutting their own triangles, distribute the pages and scissors. Otherwise, distribute sets of triangles to students. Once students have cut out the shapes, have them count to verify that they have 20 triangles in the color quantities needed (see *Management 3*).
2. Draw a line on the overhead. Place a triangle on the overhead so that its long edge is on the line. Tell students to take one of their own triangles and orient it in front of them in the same way as the one on the overhead. Instruct them to flip (reflect) their triangles across the long edge—where the line is on the overhead.
3. Invite one student to come up and show the class what he/she did with his/her triangle. See if everyone agrees that this is how the shape should have been flipped (reflected). Repeat this process with turns (rotations) and slides (translations).
4. Tell students that they will be using the triangles they cut out to make quilt squares. They will place the triangles in the quilt squares by performing transformations on the triangles according to your instructions.

SHAPES, SOLIDS, AND MORE   © 2009 AIMS Education Foundation

5. Distribute the first quilt square page and a glue stick to each student. Have them read the description at the top of the page and select the number of triangles they need (four triangles in each of two colors).
6. Read the instructions for putting the triangles on the slides (translations) quilt square. When students are done, have them compare their squares and identify any patterns or features they notice in the slides (translations) quilt square.
7. Repeat this process for the remaining two quilt squares. At the end, compare and contrast the three different patterns.

Connecting Learning
1. What is a turn (rotation)? What does it look like when you turn (rotate) a shape a quarter turn?
2. What is a flip (reflection)? What does it look like when you flip (reflect) a shape along one of its edges?
3. What is a slide (translation)? What does it look like when you slide (translate) a shape?
4. How were you able to decide where to place the shapes on your quilt squares?
5. Did you ever put a shape in the wrong place? Why? [didn't listen to the directions, flipped the wrong way, turned from the wrong corner, etc.]
6. What changed about the shapes as we put them on the quilt squares? [orientation and/or location] What stayed the same? [color, size, shape]
7. How are the three quilt squares alike? How are they different? Why?
8. Which quilt square is your favorite? Why?
9. If you were to design your own quilt square using transformations, which would you most like to use—slides (translations), flips (reflections), or turns (rotations)?

Extensions
1. Give students additional triangles and have them create their own quilt square designs. Challenge them to describe how to place the triangles in the square using transformations.
2. Explore symmetry in the quilt squares. Divide the squares students made into those with a line of symmetry and those without a line of symmetry.

Solutions
The pattern for each of the three quilt squares is shown here.

Slides (Translations) Quilt Square

Flips (Reflections) Quilt Square

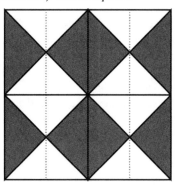

Turns (Rotations) Quilt Square

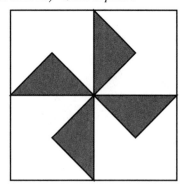

* Reprinted with permission from *Principles and Standards for School Mathematics*, 2000 by the National Council of Teachers of Mathematics. All rights reserved.

Quilted Transformations

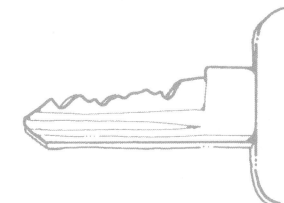

Key Question

How can you use transformations to make quilt squares?

Learning Goal

Students will:

apply transformations to shapes as directed by the teacher to create quilt squares.

Quilted Transformations

The instructions that follow can be read verbatim or they can be modified to suit your students' needs and abilities. Words in (parentheses) indicate optional vocabulary that may or may not be appropriate for your students.

Slides (Translations) Quilt Square Instructions
1. Place a triangle on the quilt square in the space indicated by the dashed lines. Glue it in place.
2. Place a second triangle of the same color on top of the one you just glued down. Slide (translate) it to the right so that it is in the second column. Glue it down.
3. Place a third triangle of the same color on top of the first triangle you glued down. Slide (translate) it to the right so that it is in the third column. Glue it down.
4. Place a fourth triangle of the same color on top of the first triangle you glued down. Slide (translate) it to the right so that it is in the fourth column. Glue it down.
5. Place a triangle of a different color in the bottom right corner of the square so that it is in the fourth column. The long edge of the triangle should touch the right edge of the square, and the bottom right corner of the triangle should touch the bottom right corner of the square. Glue it down.
6. Place a second triangle of the same color on top of the one you just glued down. Slide (translate) it to the left so that it is in the third column. Glue it down.
7. Place a third triangle of the same color on top of the first triangle you glued down. Slide (translate) it to the left so that it is in the second column. Glue it down.
8. Place a fourth triangle of the same color on top of the first triangle you glued down. Slide (translate) it to the left so that it is in the first column. Glue it down.
9. Describe your finished quilt square. What patterns do you see? What things do you notice? [The white space is small triangles and parallelograms; you could add more shapes to the quilt square using slides across the bottom, top, and center; etc.]
10. How were the slides (translations) you used different from one time to the next? [The first time you were sliding the triangle one space, the next time two spaces, and then three spaces.] Could they have been done differently? [Each time you could have put the triangle on the one you had just glued down and slid (translated) it just one space.]

Flips (Reflections) Quilt Square Instructions
1. Place a triangle on the quilt square in the space indicated by the dashed lines. Glue it in place.
2. Place another triangle on top of the first triangle. Flip (reflect) it across the dotted line. Where is your triangle? It should be in the upper left square with its long edge on the center vertical line of the quilt square. Once you have it in the right place, glue it down.
3. Place a third triangle on top of the triangle you just glued down. Flip (reflect) it across the center line. Where is your triangle? It should be in the top right square with its long edge on the center vertical line of the quilt square. Once you have it in the right place, glue it down.
4. Place another triangle on top of the triangle you just glued down. Flip (reflect) it across the dotted line. Where is your triangle? It should be in the upper right square with its long edge on the right edge of the quilt square. Once you have it in the right place, glue it down.
5. Place a fifth triangle on top of the first triangle you glued down. Flip (reflect) it down across the center horizontal line. Where is your triangle? It should be in the lower left square with its long edge on the left edge of the quilt square. Once you have it in the right place, glue it down.
6. Repeat this process with the remaining three triangles. Place each of them on top of one of the triangles in the top half of the square and flip (reflect) them down across the center horizontal line.
7. Describe the patterns you see in the quilt square. What things do you notice? [The triangles make a bowtie pattern; the two sets of triangles that are back to back form squares; the empty spaces in the middle of the quilt square are squares; etc.]
8. What if you flipped your quilt square pattern to the right along the right edge of the square? How would it look? [It would look the same.] What if you flipped it down along the bottom edge of the square? [It would look the same.]

Turns (Rotations) Quilt Square Instructions
1. Place a triangle on the quilt square in the space indicated by the dashed lines. Glue it in place.
2. Place another triangle on top of the first triangle. Turn (rotate) it to the right (clockwise) a quarter turn (90°) from the bottom left corner. Where is your triangle? It should be in the lower right square with its long edge on the center horizontal line. Once you have it in the right place, glue it down.

3. Place another triangle on top of the first triangle you glued down. Turn (rotate) it to the right (clockwise) a half turn (180°) from the bottom left corner. Where is your triangle? It should be in the lower left square with its long edge on the center vertical line. Once you have it in the right place, glue it down.
4. Take a fourth triangle and put it on top of the first triangle you glued down. Turn (rotate) it to the right (clockwise) a three-quarter turn (270°) from the bottom left corner. Where is your triangle? It should be in the upper left square with its long edge on the center horizontal line. Once you have it in the right place, glue it down.
5. Describe the patterns you see in the quilt square. What things do you notice? [It looks like a pinwheel; every time you turn it a quarter turn, it looks the same; etc.]
6. What if you turned (rotated) your quilt square pattern a quarter turn (90°)? How would it look? [It would look the same.] What if you rotated it a half turn (180°)? ...a three-quarter turn (270°)? [It would look the same.]

Make copies of this page on multiple colors of paper. Each student needs 20 triangles.

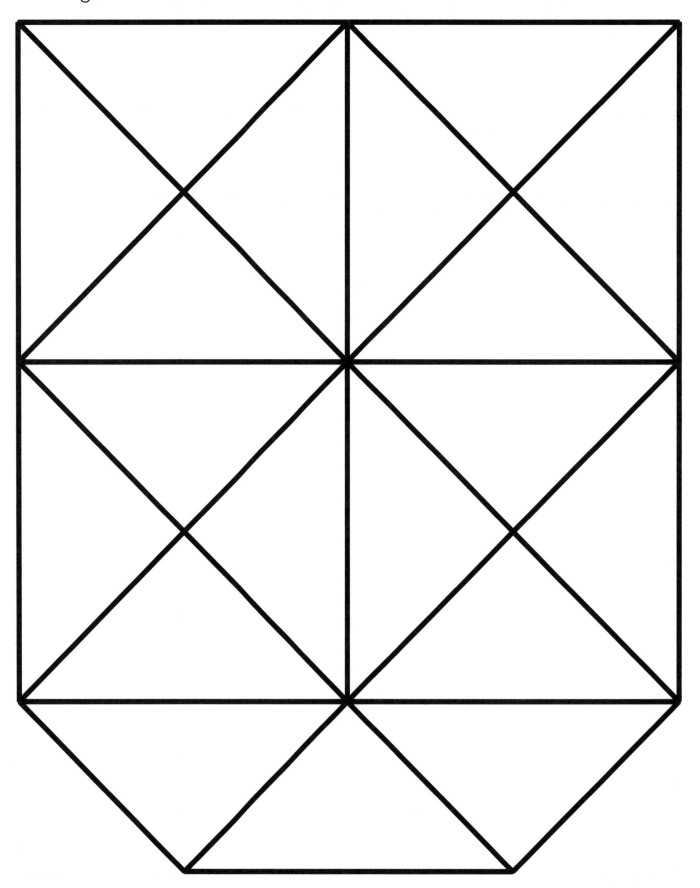

SHAPES, SOLIDS, AND MORE 264 © 2009 AIMS Education Foundation

Quilted Transformations

Slides (Translations)

This is your slides (translations) quilt square. You need four triangles of one color and four of a second color.

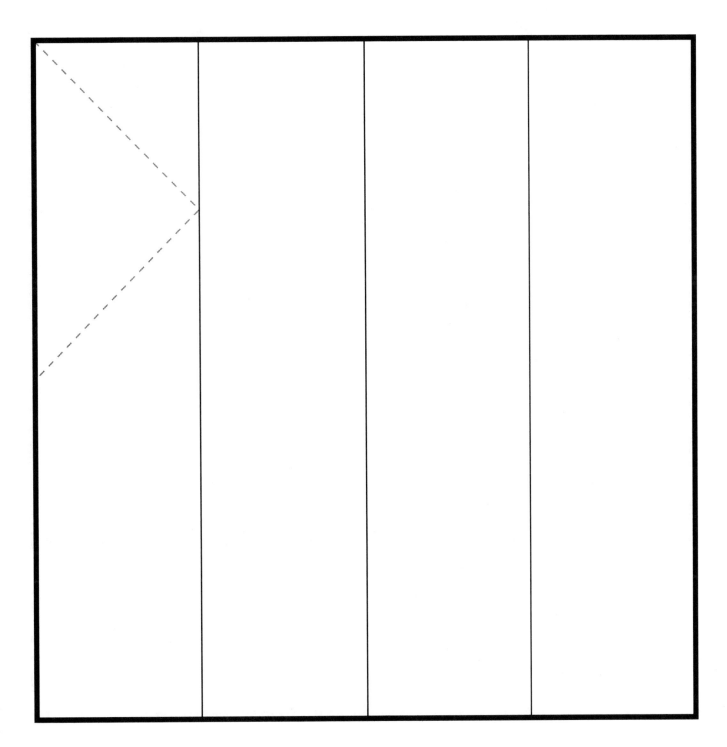

SHAPES, SOLIDS, AND MORE © 2009 AIMS Education Foundation

Quilted Transformations

Flips (Reflections)

This is your flips (reflections) quilt square. You need eight triangles that are all the same color.

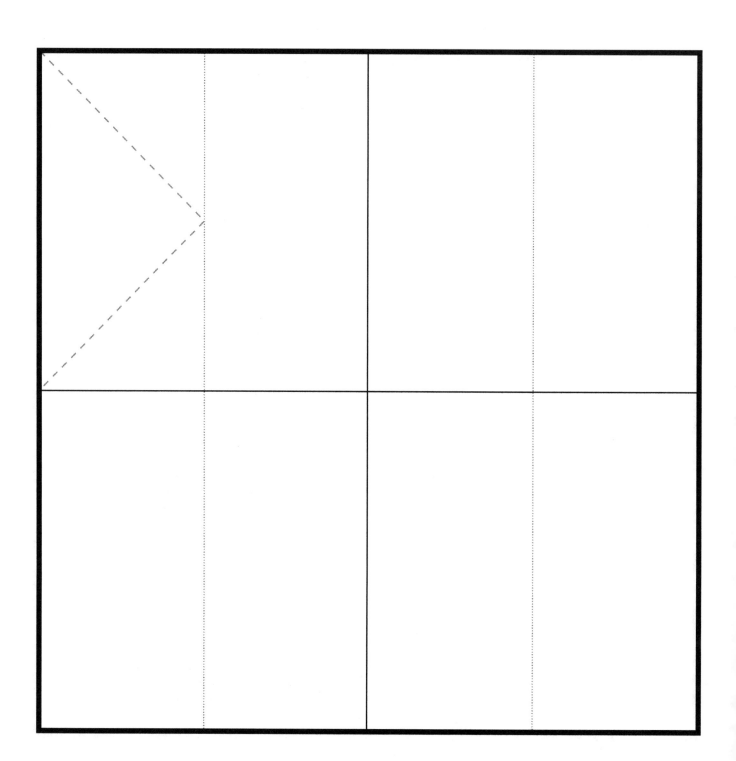

Quilted Transformations

Turns (Rotations)

This is your turns (rotations) quilt square. You need four triangles that are all the same color.

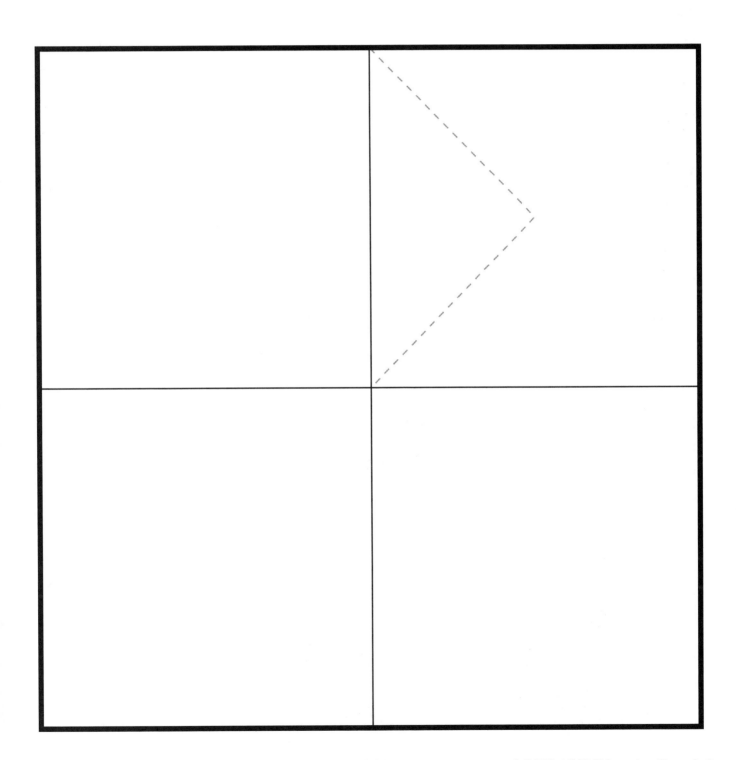

SHAPES, SOLIDS, AND MORE

Connecting Learning

1. What is a turn (rotation)? What does it look like when you turn (rotate) a shape a quarter turn?

2. What is a flip (reflection)? What does it look like when you flip (reflect) a shape along one of its edges?

3. What is a slide (translation)? What does it look like when you slide (translate) a shape?

4. How were you able to decide where to place the shapes on your quilt squares?

5. Did you ever put a shape in the wrong place? Why?

Quilted Transformations

Connecting Learning

6. What changed about the shapes as we put them on the quilt squares? What stayed the same?

7. How are the three quilt squares alike? How are they different? Why?

8. Which quilt square is your favorite? Why?

9. If you were to design your own quilt square using transformations, which would you most like to use—slides (translations), flips (reflections), or turns (rotations)?

Topic
Transformations

Key Question
What objects in the pictures have been slid (translated), flipped (reflected), and/or turned (rotated)?

Learning Goals
Students will:
- look at sets of two pictures showing the same scene; and
- identify the changes from one picture to the next by finding the slides (translations), flips (reflections), and/or turns (rotations) that have been made.

Guiding Document
NCTM Standards 2000*
- *Recognize and apply slides, flips, and turns*
- *Predict and describe the results of sliding, flipping, and turning two-dimensional shapes*
- *Investigate, describe, and reason about the results of subdividing, combining, and transforming shapes*

Math
Geometry
 transformations
  slides, flips, turns

Integrated Processes
Observing
Comparing and contrasting
Identifying

Materials
Colored pencils or crayons
Overhead pens—red, green, and blue
Student pages

Background Information
When students learn about slides, flips, and turns, they most commonly see them applied in the context of two- or three-dimensional geometric shapes. However, these transformations can be applied to everyday objects as well. This activity will challenge students to recognize slides, flips, and turns in the context of different real-world settings by comparing "before" and "after" pictures of the same scene. In each case, they will be challenged to identify the various transformations that have taken place from the "before" to the "after" picture.

Management
1. This activity is not designed as an introduction to transformations. It should be used as a culminating experience after students have learned about slides (translations), flips (reflections), and turns (rotations).
2. Make a transparency of each of the student pages.
3. The final student page can be done separately as an assessment or sent home as homework, if desired.

Procedure
1. Place the transparency of the first student page on the overhead. Cover the bottom picture. Ask students what they notice about the picture. Have them identify some key elements and objects they can see.
2. Cover the top picture and reveal the bottom picture. Ask students if this picture looks the same or if they notice any differences between it and the first picture.
3. After they study the picture for a few seconds, uncover the top picture so that students can see both at the same time. Ask them to identify some of the differences. Challenge them to use the language of geometry to describe the differences. [There are objects that have been slid (translated) to different positions.]
4. Explain that all the differences between these two pictures are slides (translations). Turn off the projector, and distribute the first student page and colored pencils to each student.
5. Tell students that their challenge is to use the colored pencils to identify each slide (translation) that has taken place in the bottom picture by coloring or circling each object that has been slid (translated).
6. Allow time for students to find all of the slides (translations), then turn on the overhead. Invite different students to come up one at a time and

SHAPES, SOLIDS, AND MORE © 2009 AIMS Education Foundation

show each of the slides (translations) by circling them on the bottom picture. When all of the slides (translations) have been discovered, distribute the second student page. Repeat the process for the pages that show flips (reflections) and turns (rotations).
7. Distribute the final student page. Tell students that this page shows all three kinds of transformations. Explain that they need to highlight all of the slides (translations) using a red colored pencil, all of the flips (reflections) using a blue colored pencil, and all of the turns (rotations) using a green colored pencil.
8. When students are done, have them identify the transformations using the overhead in the same way.
9. Discuss what students learned about transformations from this activity.

Connecting Learning
1. What kinds of transformations did you find in the first picture? [slides (translations)]
2. Were you able to find all of the slides (translations) that were pictured? If you missed any, why didn't you see them?
3. What kinds of transformations did you find in the other pictures? [flips (rotations), turns (reflections)]
4. Which kind of transformation was most difficult for you to identify? Why do you think this is?
5. How many of each kind of transformation did you find in the last picture? Did you find them all? How do you know?
6. What other examples of slides (translations), flips (reflections), and turns (rotations) can you think of using real-world objects?

Extension
Have students create scenes on their desks, in the classroom, or outside. Use a digital camera to photograph each scene, then allow them to make transformations to some of the objects in the scene and take another picture from the same perspective. Print out the photos and allow students to share them with other groups and their families to help reinforce the concepts of slides, flips, and turns.

Solutions
Slides (Translations)
    mouse
    chair
    abacus beads
    cup

Flips (Reflections)
    measuring cup
    spoon
    flower on calendar
    spatula

Turns (Rotations)
    arrow
    "swimming" paper
    black star
    boy

All transformations
    Slides (Translations): boy on slide, bird
    Flips (Reflections): flower on girl's shirt, pail
    Turns (Rotations): boy on merry-go-round, kite

* Reprinted with permission from *Principles and Standards for School Mathematics*, 2000 by the National Council of Teachers of Mathematics. All rights reserved.

Transformation Identification

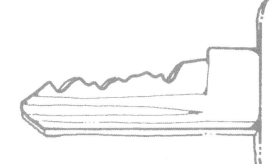

Key Question

What objects in the picture have been slid (translated), flipped (reflected), and/or turned (rotated)?

Learning Goals

Students will:

- look at sets of two pictures showing the same scene; and
- identify the changes from one picture to the next by finding the slides (translations), flips (reflections), and/or turns (rotations) that have been made.

Connecting Learning

1. What kinds of transformations did you find in the first picture?

2. Were you able to find all of the slides (translations) that were pictured? If you missed any, why didn't you see them?

3. What kinds of transformations did you find in the other pictures?

4. Which kind of transformation was most difficult for you to identify? Why do you think this is?

5. How many of each kind of transformation did you find in the last picture? Did you find them all? How do you know?

6. What other examples of slides (translations), flips (reflections), and turns (rotations) can you think of using real-world objects?

Something About Solids

Topic
Geometric solids

Key Question
How are our geometric solids alike and different?

Learning Goals
Students will:
• compare and contrast geometric solids: and
• explore different orientations, sizes, and types to discover that each solid has certain distinguishable characteristics.

Guiding Documents
Project 2061 Benchmark
• *Simple graphs can help to tell about observations.*

NCTM Standards 2000*
• *Recognize, name, build, draw, compare, and sort two- and three-dimensional shapes*
• *Describe attributes and parts of two- and three-dimensional shapes*
• *Identify, compare, and analyze attributes of two- and three-dimensional shapes and develop vocabulary to describe the attributes*

Math
Geometry
　　3-D solids
　　　　characteristics

Integrated Processes
Observing
Recording
Comparing and contrasting
Communicating

Materials
For each student:
　　Something About Solids journal

For the class:
　　sets of AIMS Geo-Solids (see *Management 3*)
　　chart paper
　　mystery bag (see *Management 1*)

Background Information
　　Students should be given the opportunity to explore and discover attributes of geometric solids. Concrete experiences afford the opportunities to direct young students to the observations of the properties of solids.

　　In this activity, students are introduced to models of geometric solids. They will compare the models to each other and to real-world examples of the solids.
　　The following list will provide some factual information about geometric solids.
• Spheres are round.
• Rectangular solids have six faces and each face is rectangular. Note: A cube is a special rectangular solid that has six faces that are all the same size.
• Cones have at least one face (called the base) and a vertex that is not on the face. It is possible to draw a straight line from any point on the edge of the base to the vertex. If the base is a circle, the cone is a circular cone—the type most people associate with the word *cone*. The base of a cone, however, can be any shape. A pyramid is a special cone in which the base is a polygon. All of the faces of a pyramid are triangles, except possibly the base.
• Cylinders have two congruent faces in parallel planes called bases. All elements joining corresponding points on the bases are parallel.

Management
1. Prior to teaching the lesson, prepare a mystery bag. Take a large brown grocery bag and fold it as illustrated. Measure six inches down from the top and cut around on that mark. Open the bag and, starting at the point of the triangle, cut out a three-inch square piece. Cut out a square piece on the opposite side.

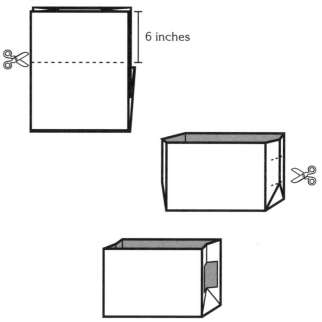

SHAPES, SOLIDS, AND MORE

2. The mystery bag is used in *Part Two*. It is used to conceal the Geo-Solids from view. Students will need to use their sense of touch and their knowledge of the properties of the solids to identify them.

3. Geo-Solids are available from AIMS (item numbers 4610 or 4612). If Geo-Solids are limited in supply, use real-world examples such as soup cans and tissue boxes.

Procedure
Part One
1. Display a sphere. Ask students to look around the room and locate examples of spheres. Ask them to go and get the spheres if possible or to stand near them.
2. Ask the class what characteristics an object has to have to be considered a sphere. Record student responses on a piece of chart paper under the label *sphere*. (Students will be coming up with an operational definition of the shape, so encourage them to be as specific as possible.) Ask them to include things such as how it moves and if it stacks.
3. Using the student-generated definition, examine the objects the students have selected to determine if they are good examples of spheres.
4. Distribute the *Something About Solids* journals and give each group of three to four students a Geo-Solid.
5. Ask groups to examine their shapes carefully, finding out if they will roll, slide, or stack and describing them in their journals.
6. Have students find other class objects that are the shape of the solids they are examining. Have them record the objects using pictures or words in their journals.
7. Before groups rotate solids, ask them to describe the shapes in their journals. Remind them to include something they discovered about the shapes as well as what it looks like.
8. Rotate the solids until each group has an opportunity to examine all of the models.
9. End with a time for students to share what they discovered about each solid. Record the attributes they used in their descriptions of each shape.

Part Two
1. Invite a student to close his/her eyes and place his/her hands in the mystery bag. Give the student one of the Geo-Solids and ask him/her to identify the mystery shape. Question the student about the attributes that helped in making the decision. Continue this process until all students have had a chance to identify a solid.
2. End with a time of discussion about how students identified the shapes without being able to use their eyes.

Connecting Learning
1. How did you go about choosing a shape in the classroom that represented each of the solids?
2. How did you identify the mystery shape without using your eyes?
3. How can you tell the difference between a cube and a rectangular solid?

* Reprinted with permission from *Principles and Standards for School Mathematics*, 2000 by the National Council of Teachers of Mathematics. All rights reserved.

Something About Solids

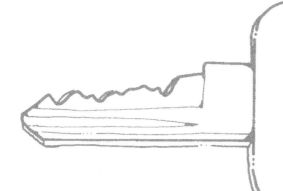

Key Question

How are our geometric solids alike and different?

Learning Goals

Students will:

- compare and contrast geometric solids: and
- explore different orientations, sizes, and types to discover that each solid has certain distinguishable characteristics.

Rectangular Solid

1. Can a rectangular solid roll? _____

2. Can a rectangular solid stack? _____

3. Can a rectangular solid slide? _____

4. What classroom items are this shape?

5. Describe a rectangular solid.
_____
_____
_____
_____

✂--------------------------------

Something about Solids Journal

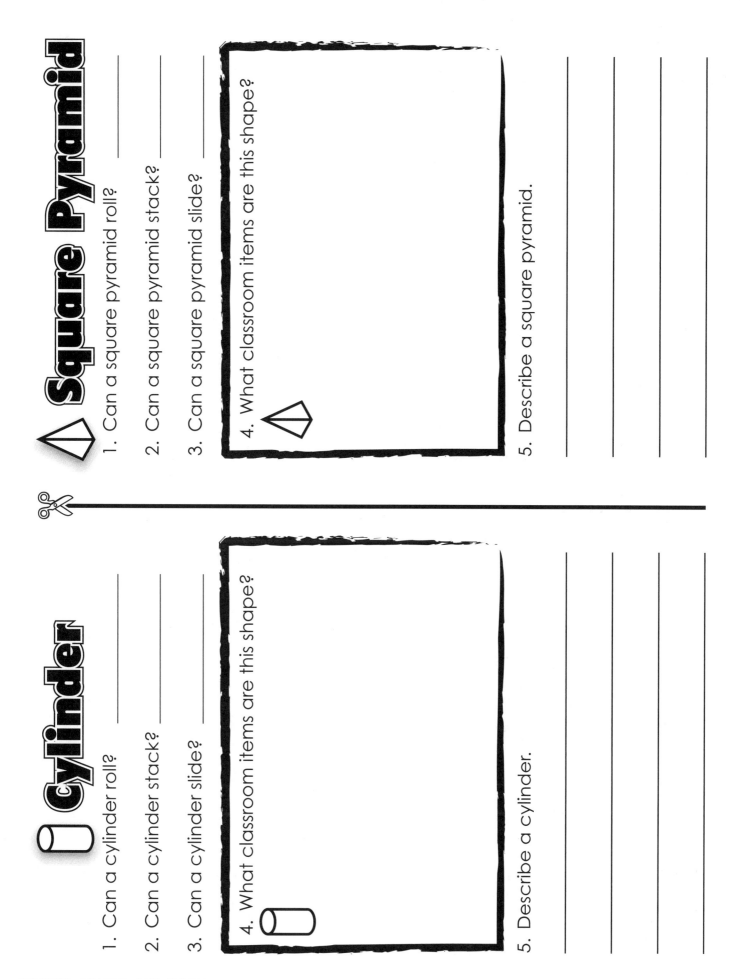

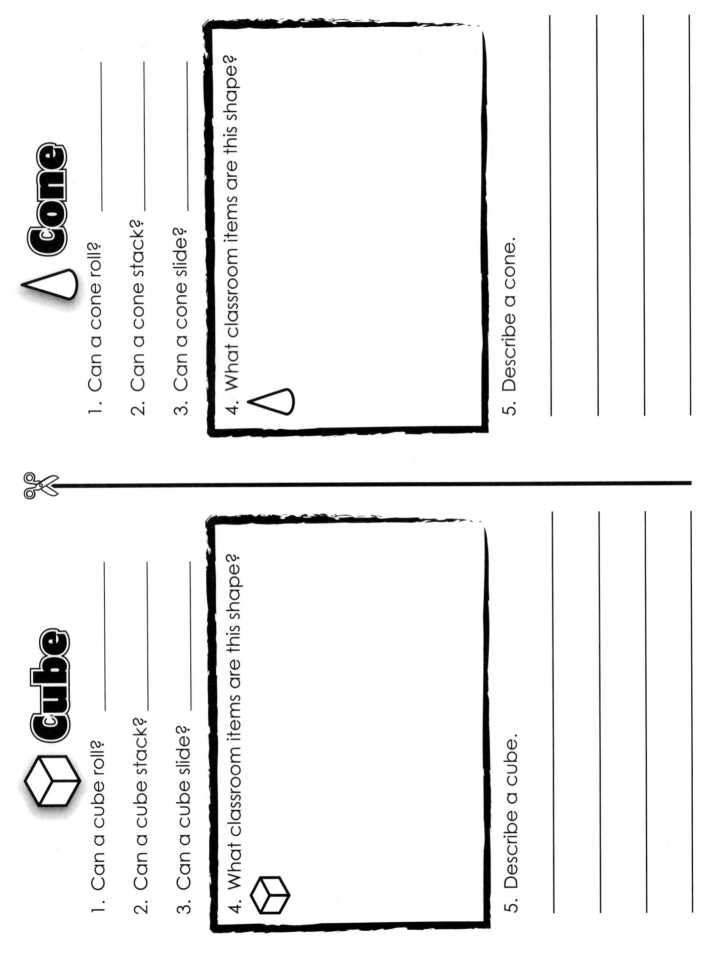

Something About Solids

Connecting Learning

1. How did you go about choosing a shape in the classroom that represented each of the solids?

2. How did you identify the mystery shape without using your eyes?

3. How can you tell the difference between a cube and a rectangular solid?

Shape Shadows

Topic
Geometric solids

Key Question
What shadow shapes do three-dimensional objects make?

Learning Goals
Students will:
- explore the shadows made by three-dimensional objects and geometric solids, and
- relate the shadow shapes to common two-dimensional shapes.

Guiding Document
NCTM Standards 2000*
- *Describe attributes and parts of two- and three-dimensional shapes.*
- *Recognize and represent shapes from different perspectives*

Math
Geometry
  shape recognition

Integrated Processes
Observing
Recording
Comparing and contrasting
Communicating

Materials
For each group of students:
  flashlight
  large piece drawing paper, minimum 3' x 1.5'
  masking tape
  real-world geometric solids (see *Management 1*)
  AIMS Geo-Solids
  four empty film canisters or wooden blocks
  crayons

For each student:
  Shape Shadows booklet (see *Management 5*)
  scissors
  glue

Background Information
Geometric activities provide some of the best experiences for children to acquire and develop spatial awareness. Through this activity, children will begin to build a connection between real-world, three-dimensional objects and the abstract, two-dimensional shapes commonly taught in the early grades. Through the exploration of the shadows cast from real-world objects and geometric solid models, the students will discover two-dimensional shapes.

Management
1. Gather a collection of real-world geometric solids such as soup cans, tissue boxes, balls, dice, party hats, and cereal boxes.
2. This activity is nicely managed at a station with a group of four to six students.
3. It is important to hold the flashlight directly overhead so the shapes aren't distorted. Only the model should be rotated when casting the shadows.
4. Prior to this lesson, give the students time to investigate a set of geometric solids using flashlights.
5. Duplicate and assemble a *Shape Shadows* booklet for each student.
6. Geo-Solids are available from AIMS (item numbers 4610 and 4612).

Procedure
Part One
1. Tell the students that they will be making shape shadows. Demonstrate how to place a geometric solid or real-world object on top of a film canister or wooden block. Place the canister and object on top of a piece of paper. Using masking tape, secure a flashlight to the edge of a chair, pointing downward. Show students how to place the object and canister directly under the flashlight. Ask a student to trace the shadow that is cast on a piece of paper.

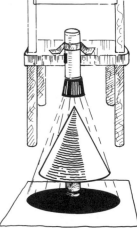

SHAPES, SOLIDS, AND MORE   287   © 2009 AIMS Education Foundation

2. Give the students large pieces of drawing paper, and challenge them to record shadows of at least two real-world objects they think are similiar to each geometric solid they have at the center. For example, the students might draw the shadows of a tissue box and a block to correspond to the cube, a snowcone holder and a party hat to correspond to the cone, etc. Tell them to also draw the shadows of the geometric solids.
3. Once the groups have finished tracing the shadows, ask them to trade their tracings and the objects they used to make the shadows with another group. Challenge each group to match the shadow drawings with the real-world objects used to make the shadows.

Part Two
1. Give each student a *Shape Shadows* booklet.
2. Using the same shadow-making technique used in *Part One*, challenge the students to find solids that can cast a shadow of the shape shown on each page of their booklet. Ask the students to draw an example of a real-world object for each shape shadow as well.

Connecting Learning
1. Show a shadow that can be made from a sphere. ...from a cube. ...from a cylinder. ...from a cone. ...from a rectangular solid. [*sphere*—circle; *cylinder*—circle, square, rectangle; *rectangular solid*—square, rectangle; *cube*—square; *cone*—circle, triangle]
2. What did you notice about the shadows and the objects that made them? [You can find some of the shapes on the objects.]

* Reprinted with permission from *Principles and Standards for School Mathematics*, 2000 by the National Council of Teachers of Mathematics. All rights reserved.

Shape Shadows

Key Question

What shadow shapes do three-dimensional objects make?

Learning Goals

Students will:

- explore the shadows made by three-dimensional objects and geometric solids, and
- relate the shadow shapes to common two-dimensional shapes.

SHAPE SHADOWS

What can make this shape shadow?

What can make this shape shadow?

What can make this shape shadow?

Shape Shadows

Connecting Learning

1. Show a shadow that can be made from a sphere. …from a cube. …from a cylinder. …from a cone. …from a rectangular solid.

2. What did you notice about the shadows and the objects that made them?

Sorting Out Solids

Topic
Geometric solids

Key Question
Where are the faces, edges, and vertices on a three-dimensional solid?

Learning Goals
Students will:
- identify the faces, edges, and vertices on three-dimensional solids;
- sort a given set of three-dimensional solids by the number of faces, edges, or vertices they have; and
- create three line plots based on the number of faces, edges, or vertices found on the given set of three-dimensional solids.

Guiding Documents
Project 2061 Benchmarks
- *Numbers and shapes can be used to tell about things.*
- *Use numerical data in describing and comparing objects and events.*

NCTM Standards 2000*
- *Recognize, name, build, draw, compare, and sort two- and three-dimensional shapes*
- *Describe attributes and parts of two- and three-dimensional shapes*
- *Identify, compare, and analyze attributes of two- and three-dimensional shapes and develop vocabulary to describe the attributes*

Math
Geometry
    3-D solids
        faces, edges, vertices
Graphing
    line plot

Integrated Processes
Observing
Classifying
Collecting and recording data
Comparing and contrasting
Interpreting data

Materials
For each group:
    1 set of 3-D models (see *Management 1*)
    1 meter of adding machine tape (see *Management 2*)
    crayons
    scissors
    glue

For the class:
    yellow, red, and blue sticky dots, one of each per group
    chart paper
    sticky tacky

Background Information
This activity provides an opportunity for students to compare and contrast and explore the properties of select three-dimensional solids tactilely. At this level students should understand that:
- a face is a flat surface;
- an edge is where two flat surfaces meet;
- cylinders have curved edges, or rims, two faces, and a curved surface; and
- a cone has one vertex, one face, and one curved surface.

In the activity, students will sort their sets of three-dimensional solids by number of faces they have, the number of edges, and number of vertices. They will create simple line plots to display their findings. A line plot is a simple form of graphing that shows the distribution of data horizontally. It will allow students to see common clustered areas and gaps where no solid fits the description.

Management
1. Manufactured sets of Geo-Solids can be purchased from AIMS (item numbers 4610 or 4612). If models are not available for each group of students, real-world examples of the three-dimensional solids can be used. Having a food drive prior to starting your three-dimensional geometry unit will provide your class with a variety of real-world cylinders, cubes, and rectangular solids. It will also allow students to participate in a community service project that will help needy families.
2. Prepare adding machine tape number lines. Write the numbers 0 through 12 on the adding machine tape. Place the numbers about five centimeters (two inches) apart.
3. Groups of three or four students work well for this activity.

SHAPES, SOLIDS, AND MORE

Procedure

1. Display the rectangular solid model so that the entire class can clearly see it. Ask if anyone in class can identify it. (Naming the solid is not the focus of this activity.) Tell the class that it is called a rectangular solid. Point to one of the flat surfaces and explain to the class that all of the flat surfaces are called faces.
2. Ask the students how many faces they think a rectangular solid has.
3. Give each group of students a rectangular solid and one yellow sticky dot. Have the students place the sticky dot on one of the flat surfaces of their rectangular solids. Ask them to draw a smiley face on the dot. Explain that this will remind them that they are counting "faces" and that it will mark where they started counting. Place one yellow sticky dot on a piece of chart paper and beside it write the word *faces*. Discuss how many faces a rectangular solid has.
4. Display the rest of the 3-D models. Ask students to predict how many faces are on each of the solids.
5. Tape one of the number lines constructed in *Management 2* to the board. Write the word *faces* below the number line. Ask the students how many faces are on a rectangular solid. Attach the rectangular solid to the board above the number six on the number line using sticky tacky.
6. Distribute a set of 3-D models and an adding machine tape number line to each group. Instruct the students to count the faces and place each model on the number line above the number they counted.
7. Give each group a line plot page, a set of 3-D drawings, scissors, and glue. Explain that they are going to make a permanent record of their results by cutting and gluing the pictures on the page to match their real graph (line plot). Be sure they write *faces* in the blank to label their number line.
8. Repeat the procedure for number of edges and vertices (corners). Use red dots to identify edges and blue dots for vertices.

Connecting Learning

1. What is a face? [a flat surface on a solid] ...an edge? [a place where two faces meet] ...a vertex? [a point or corner of a solid]
2. Which solid(s) had the most faces? [The rectangular solid and the cube each have six faces.] ...edges? [The rectangular solid and the cube each have 12 edges.] ...vertices? [The rectangular solid and the cube each have eight vertices.]
3. Which picture did you use to represent the rectangular solid? ...cylinder? ...cube?
4. Where did you put the sphere? Explain your thinking.

Solutions

3-D Solid	Faces	Edges	Vertices
Rectangular Solid	6	12	8
Sphere	0	0	0
Cone	1	1	1
Cube	6	12	8
Cylinder	2	2	0
Square Pyramid	5	8	5

* Reprinted with permission from *Principles and Standards for School Mathematics*, 2000 by the National Council of Teachers of Mathematics. All rights reserved.

Sorting Out Solids

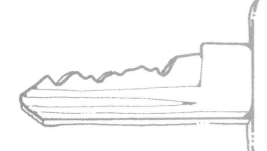

Key Question

Where are the faces, edges, and vertices on a three-dimensional solid?

Learning Goals

- identify the faces, edges, and vertices on three-dimensional solids;
- sort a given set of three-dimensional solids by the number of faces, edges, or vertices they have; and
- create three line plots based on the number of faces, edges, or vertices found on the given set of three-dimensional solids.

Sorting Out Solids

Make a record of your sorting. Glue the shape pictures above the correct numbers.

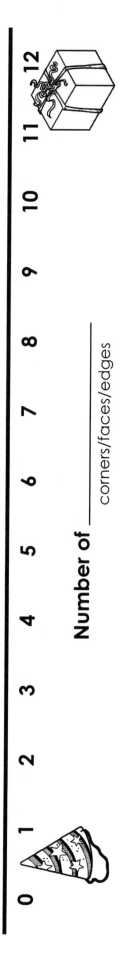

0 1 2 3 4 5 6 7 8 9 10 11 12

Number of _____ corners/faces/edges

SHAPES, SOLIDS, AND MORE

Sorting Out Solids

SHAPES, SOLIDS, AND MORE © 2009 AIMS Education Foundation

Sorting Out Solids

Connecting Learning

1. What is a face? ...an edge? ...a vertex?

2. Which solid(s) had the most faces? ...edges? ...vertices?

3. Which picture did you use to represent the rectangular solid? ...cylinder? ...cube?

4. Where did you put the sphere? Explain your thinking.

Topic
Geometric solids

Key Question
What attributes did you use to sort your solids?

Learning Goal
Students will play a game during which they will sort various solids into like groups.

Guiding Documents
Project 2061 Benchmarks
- Circles, squares, triangles, and other shapes can be found in things in nature and in things that people build.
- Shapes such as circles, squares, and triangles can be used to describe many things that can be seen.

NCTM Standards 2000*
- Recognize, name, build, draw, compare, and sort two- and three-dimensional shapes
- Describe attributes and parts of two- and three-dimensional shapes

Math
Geometry
    3-D solids

Integrated Processes
Observing
Comparing and contrasting
Sorting
Identifying
Communicating

Materials
AIMS Geo-Solids
Various real-world examples of solids
    (see *Management 3*)
Paper grocery bags or small boxes for sorting
Labels for the bags or boxes

Background Information
    Young children need repeated experiences in order to develop and extend their understandings of geometry. Using geometric solids from the real world, as well as models of geometric solids, provides opportunities for students to explore the characteristics and properties of both. As they sort and sort again, they will begin to notice similarities and differences in the various shapes.
    This activity is designed to engage students in a physical game while applying the sorting of solids by similar attributes. They will play a relay-type game in which they run to a given area, sort a solid, and return to touch the next member of the team. The students can play the game over and over as a review, and decide on multiple ways in which to sort the objects for the game.

Management
1. It is assumed that students have had repeated experiences sorting and identifying solids before playing the game.
2. A large and varied collection of geometric solid shapes should be gathered prior to the activity. Both the AIMS Geo-Solids (item numbers 4610 or 4612) and real-world examples of those solids should be used. Try to gather a variety of sizes of the same shape. You will need at least one object for every person in the class.
3. Use the page provided to label the boxes or bags that will be used as sorting vessels. The first few times the game is played, use the common names for the solids. (Cone, sphere, cube, rectangular prism, etc.) When students are ready for a change in the rules, they may suggest other attributes by which the shapes can be sorted. (Straight sides, corners, straight lines, points, etc.) Use the blank spaces provided to make your own labels for this portion of the activity.
4. Each team should have its own set of solids and sorting containers. The number of solids in each area should be equivalent to the number of students on each team.

Procedure
1. Divide the students into two to four teams. There should be at least four students per team.
2. Review the names of the solids students will be sorting and have students share the characteristics of these solids. [spheres are round, cubes have square sides, cones have a pointed end and an end that's a circle, etc.]
3. Show the teams the collections of geometric solids and the areas where they will sort the solids into the containers.

SHAPES, SOLIDS, AND MORE   © 2009 AIMS Education Foundation

4. Have one student model reaching into their collection bag, taking a solid, running to a sorting area, sorting it into the correct container, and running to touch the next person on his/her team.
5. Explain that when each member of the team has run, sorted an object, and returned to the line, that team is finished. The first team to finish with all solids sorted correctly is the winner.
6. After playing the game several times with the various sorting labels, take the objects in the bags and organize them into a real graph. Discuss what students notice about the objects they sorted.

Connecting Learning
1. How did you decide which container to sort your solid into?
2. Which solid did we sort the most of? Why might this be the case? [There are more examples of that type of solid in the real world.]
3. Did each group start with the same number of each solid? Explain. [No.]
4. Why do you think we got more of one type than another?
5. Are there any shapes we have studied that were not in the collection? Why or why not?
6. Can you think of other ways we could have sorted the solids? [straight sides, curved sides, number of corners, etc.]
7. What solids would you like to add to the collection for the next time we play the game?

Extension
Have students search their homes and the playground for interesting examples of spheres, cylinders, prisms, cubes, and cones. After collecting the solids, the students can compare and sort the solids in various ways. They will begin to notice similarities and differences in the various shapes as they sort and sort again.

* Reprinted with permission from *Principles and Standards for School Mathematics*, 2000 by the National Council of Teachers of mathematics. All rights reserved.

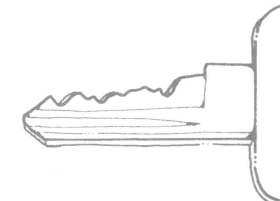

Key Question

What attributes did you use to sort your solids?

Learning Goal

Students will:

play a game during which they will sort various solids into like groups.

Cube	
Rectangular Prism	
Sphere	
Cylinder	
Cone	

Connecting Learning

1. How did you decide which container to sort your solid into?

2. Which solid did we sort the most of? Why might this be the case?

3. Did each group start with the same number of each solid? Explain.

4. Why do you think we got more of one type than another?

5. Are there any shapes we have studied that were not in the collection? Why or why not?

Connecting Learning

6. Can you think of other ways we could have sorted the solids?

7. What solids would you like to add to the collection for the next time we play the game?

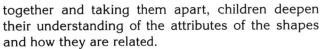
Construction Zone

Topic
Geometric solids

Key Question
How can we combine a set of three-dimensional solids to make different structures?

Learning Goals
Students will:
• identify the geometric solids that make up specific structures, and
• use problem-solving strategies to assemble the geometric solid combinations.

Guiding Document
NCTM Standards 2000*
• *Recognize, name, build, draw, compare, and sort two- and three-dimensional shapes*
• *Investigate and predict the results of putting together and taking apart two- and three-dimensional shapes*
• *Build new mathematical knowledge through problem solving*
• *Apply and adapt a variety of appropriate strategies to solve problems*

Math
Geometry
   3-D solids
Problem solving

Integrated Processes
Observing
Identifying
Recording

Problem-Solving Strategies
Use manipulatives
Guess and check

Materials
Construction Cards
AIMS Geo-Solids (see *Management 1*)

Background Information
An important geometric idea for young children to explore is that shapes can be combined or subdivided to make other shapes. For example, they should discover that two cubes will combine to form a rectangular prism. By putting shapes together and taking them apart, children deepen their understanding of the attributes of the shapes and how they are related.

In this activity, students are asked to look at a set of two-dimensional drawings of three-dimensional shapes that have been combined to form new shapes. They are to look carefully at the parts of the shapes and then determine what they will need to build a model that matches the picture. Moving back and forth between 3-D objects and their 2-D representations will help students better understand the characteristics of common 3-D shapes.

Management
1. Each pair of students will need a set of geometric solids. Geo-Solids are available from AIMS (item numbers 4610 or 4612). If geometric solids are not available, real-world examples of the solids can be substituted. For example, a soup can makes a great cylinder, a rubber ball is a sphere, and a tissue box is a rectangular solid.
2. While it is not necessary for students to identify the faces, edges, and vertices of each solid in order to combine them to make new shapes, this activity will provide an opportunity for you to introduce or reinforce geometric terms as they work with the shapes.
3. This activity can be done with the entire class or at a learning center. Each group can get the same cards or you may choose to cut them apart and give different cards to different students based on ability.

Procedure
1. Display a cube. Ask students to name the solid. Have them identify other similarly shaped objects in the classroom. Question the students about whether the cube can easily stack or combine with additional cubes or other three-dimensional shapes.
2. Display the other geometric solids. Explain that they are also geometric solids. Question the students about which shapes they think will stack well and which will not stack well.
3. Ask students to identify two solids that might combine to make a castle tower. Invite a student to build a tower with the two solids of his/her choice. Discuss how the problem-solving strategy of guessing and checking can be used to see if they were right. Question students about any other solids that would combine to make a tower.

SHAPES, SOLIDS, AND MORE

4. Tell the students that they will be exploring other combinations of shapes. Explain that each pair of students will get a set of geometric solids and a set of construction cards. Display a construction card. Instruct the students to decide which of the shapes at the bottom of the page they will need to use to build the structure.
5. Distribute materials and allow time for students to work.
6. End with a discussion about the strategies they used in constructing models of the pictured shapes.

Connecting Learning
1. What are some things that you observed about the solids?
2. How are the cube and rectangular solid the same? How are they different?
3. How did you decide which solids you would need to build the shape in the picture? Did you choose the right solids?
4. Why do you think we used the pieces we chose to build a model of what was in the pictures?
5. Which structures were the easiest to build? ...the hardest? Why?

* Reprinted with permission from *Principles and Standards for School Mathematics*, 2000 by the National Council of Teachers of Mathematics. All rights reserved.

Construction Zone

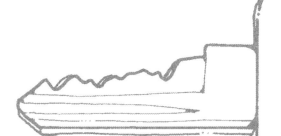

Key Question

How can we combine a set of three-dimensional solids to make different structures?

Learning Goals

- identify the geometric solids that make up specific structures, and
- use problem-solving strategies to assemble the geometric solid combinations.

construction zone

Which of these shapes will you use?

Which of these shapes will you use?

SHAPES, SOLIDS, AND MORE

Construction Zone

Which of these shapes will you use?

Which of these shapes will you use?

SHAPES, SOLIDS, AND MORE   309   © 2009 AIMS Education Foundation

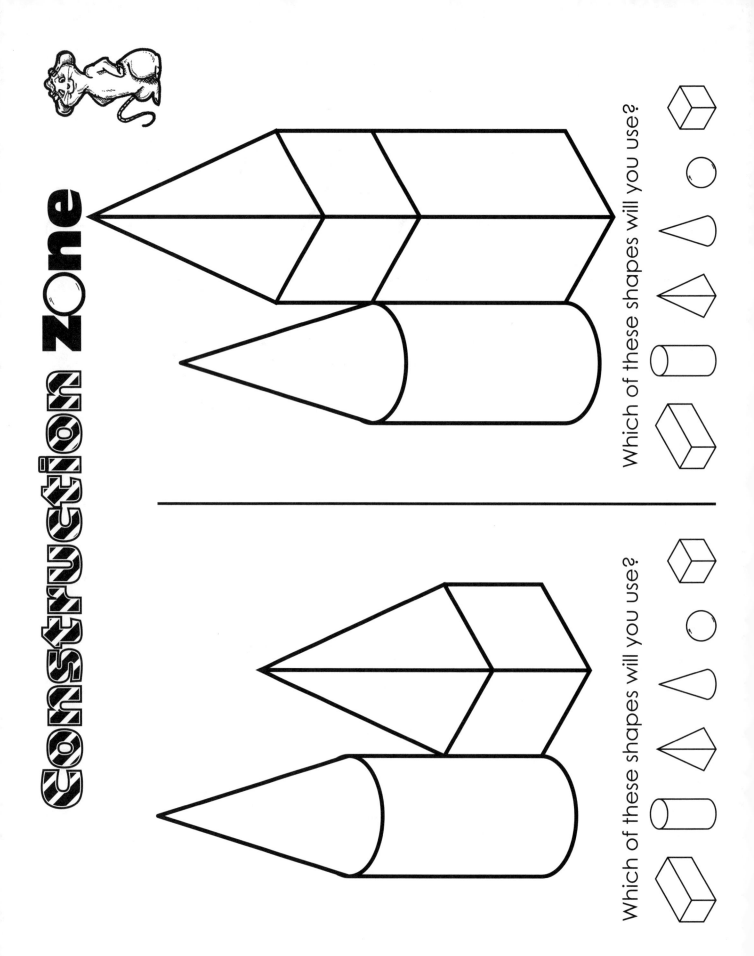

Construction Zone

Connecting Learning

1. What are some things that you observed about the solids?

2. How are the cube and rectangular solid the same? How are they different?

3. How did you decide which solids you would need to build the shape in the picture? Did you choose the right solids?

4. Why do you think we used the pieces we chose to build a model of what was in the pictures?

5. Which structures were the easiest to build? ...the hardest? Why?

Cool Castles

Topic
Geometric solids

Key Questions
1. Which geometric solids stack well?
2. How can we combine three-dimensional solids to make a castle?

Learning Goals
Students will:
• combine geometric solids made of ice to form a castle, and
• identify the geometric solids used by others to make their "cool" castles.

Guiding Documents
Project 2061 Benchmark
• *Shapes such as circles, squares, and triangles can be used to describe many things that can be seen.*

NCTM Standards 2000*
• *Recognize, name, build, draw, compare, and sort two- and three-dimensional shapes*
• *Investigate and predict the results of putting together and taking apart two- and three-dimensional shapes*
• *Build new mathematical knowledge through problem solving*
• *Apply and adapt a variety of appropriate strategies to solve problems*

Math
Geometry
   3-D solids
      composition
      decomposition

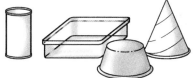

Integrated Processes
Observing
Comparing and contrasting
Identifying
Recording

Materials
For each group of students:
  a variety of containers (see *Management 1*)
  pie tin (see *Management 3*)
  salt (see *Management 5*)
  paper cup, 3 oz
  student page, one per student

For the class:
  pictures of castles
  several pitchers of prepared drink mix
  set of geometric solids (see *Management 4*)
  non-stick cooking spray

Background Information
Geometry is easily seen in the architecture of our schools, homes, and office buildings. When planning these structures, shapes are often combined or subdivided to create new shapes. Builders need to have an understanding of geometry. They have to consider which shapes stack well, which shapes efficiently and aesthetically combine with other shapes, etc.

It is important for students to explore the idea that shapes can be combined or subdivided to make other shapes. For example, they should discover that two cubes combine to form a rectangular prism. By putting shapes together and taking them apart, children deepen their understanding of the attributes of the shapes and how they are related. As students progress through school, they will often depend on their knowledge of geometry to help them with class projects.

In this activity, students will identify three-dimensional shapes used in castles and combine some three-dimensional shapes to create a "cool" castle of their own.

Management
1. Prior to the lesson, collect a variety of containers. Containers must be suitable to put in the freezer and have the shape of a geometric solid. For example, small juice cans and film canisters make nice cylinders, milk cartons can be used to make cubes and rectangular prisms, and snow cone cups make nice cones. Spraying containers with a non-stick cooking spray will help when removing the ice shapes. Each group will need several containers. Lining boxes with aluminum foil will also allow you to make more shapes.
2. Power-Solids, three-dimensional geometric solids, are available from AIMS (item number 4600).
3. Make arrangements to use a large school freezer.
4. Gather one pie tin or cookie sheet large enough to hold a "cool" castle for each group.
5. Each group will need one to two tablespoons of salt in a three-ounce paper cup.

SHAPES, SOLIDS, AND MORE   © 2009 AIMS Education Foundation

Procedure
1. Display a set of geometric solids. Ask students to name each of the solids. Question the students as to whether the three-dimensional shapes can easily be stacked or combined with other three-dimensional shapes.
2. Have students identify two shapes that might combine to make a house. Invite a student to build a tower with the two shapes of his/her choice.
3. Tell the students that they will be exploring other combinations of shapes. Inform them that they will be making "cool" castles and that they need to select different containers in which their group will freeze some water to make the ice to build the castles.
4. Have them label the containers by writing their group name on masking tape.
5. Direct them to partially fill each container with water, leaving room for the expansion of the ice. Discuss what shape and thickness the ice will be from each container.
6. Distribute materials and allow time for students to work.
7. Freeze the water.
8. After the water has frozen, help students remove the ice from the containers. Distribute the cup of salt to each group.
9. Demonstrate how to join the ice forms by lightly sprinkling salt on one surface of the ice. As soon as some liquid appears, place another piece of ice on top. The water will begin to freeze thus joining the two pieces.
10. Once all the castles have been built, discuss the new shapes that were formed. Discuss which shapes were easiest to stack.
11. Direct students to draw pictures of their ice castles, identifying each three-dimensional shape used in the construction.
12. End with a discussion about which solids stack well and the various combinations used.

Connecting Learning
1. What solid shapes did you use in your castle?
2. Were some solids easier to build with than others? Which ones? Why?
3. What shapes did you and your classmates use most often? Why?
4. Where there any shapes that you did not use? Why?

* Reprinted with permission from *Principles and Standards for School Mathematics,* 2000 by the National Council of Teachers of Mathematics. All rights reserved.

Cool Castles

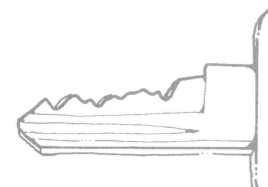

Key Questions

1. Which geometric solids stack well?
2. How can we combine three-dimensional solids to make a castle?

Learning Goals

- combine geometric solids made of ice to form a castle, and
- identify the geometric solids used by others to make their "cool" castles.

COOL CASTLES
Draw a picture of your ice castle.

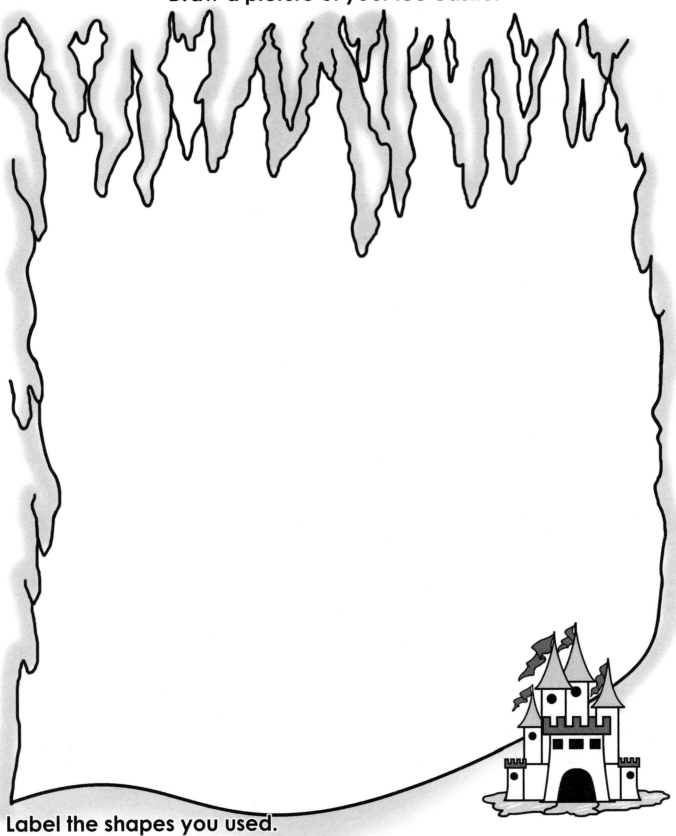

Label the shapes you used.

SHAPES, SOLIDS, AND MORE

Cool Castles

Connecting Learning

1. What solid shapes did you use in your castle?

2. Were some solids easier to build with than others? Which ones? Why?

3. What shapes did you and your classmates use most often? Why?

4. Where there any shapes that you did not use? Why?

3-D Designs

Topic
3-D solids

Key Question
What kind of three-dimensional structures can we make?

Learning Goals
Students will:
- build a cube using chenille stems and straws, and
- use this figure as a base to explore the building process further.

Guiding Documents
Project 2061 Benchmarks
- *Assemble, describe, take apart and reassemble constructions using interlocking block, erectors sets, and the like.*
- *Circles, squares, triangles, and other shapes can be found in things in nature and in things that people build.*
- *When trying to build something or get something to work better, it usually helps to follow directions if there are any or to ask someone who has done it before for suggestions.*
- *Several steps are usually involved in making things.*

*NCTM Standards 2000**
- *Recognize, name, build, draw, compare, and sort two- and three-dimensional shapes*
- *Describe attributes and parts of two- and three-dimensional shapes*
- *Investigate and predict the results of putting together and taking apart two- and three-dimensional shapes*
- *Build and draw geometric objects*

Math
Geometry
    3-D solids

Integrated Processes
Observing
Recording
Communicating
Applying

Materials
Part One (for each student):
    12 drinking straws
    16 1-inch pieces of chenille stems

Part Two:
    additional supply of drinking straws
    additional supply of 1-inch pieces of chenille stems
    paper

Part Three (for each student):
    12 cotton swabs (see *Management 5*)
    recloseable plastic bag

Part Three (for the teacher):
    rubber cement

Background Information
Through the process of building, taking apart, changing the design, and improving the techniques of construction, children begin to build an understanding of how their world works. Architectural designs are based on geometric shapes and the relationships between those shapes. While exploring three-dimensional shapes through the building process, students will gain spatial awareness and a familiarity with some of the attributes of geometric solids.

Management
1. This activity is designed in three parts. The first part is quite directed to teach a process of building. The second part is designed as very open ended to encourage creativity from each student. Through the trial and error process of building additional structures, the students will begin to construct their own knowledge about the strength of materials, need for support, etc.
2. You will need to precut the chenille stems into one-inch lengths.
3. Use straws with a narrow diameter so the chenille stems will fit snugly.
4. It is suggested that you teach the process of building the cube in small groups. Once the students learn the process, they will be able to easily build the three-dimensional figures independently.

SHAPES, SOLIDS, AND MORE

5. For *Part Three*, dip both ends of each cotton swab into rubber cement, place on foil or waxed paper, and let dry for approximately 15 minutes. The cotton swabs will remain "tacky" for use over an extended period of time. Store in a recloseable plastic bag. Twelve is a minimum number necessary to supply to each student. You may want to offer a larger supply for extended exploration.

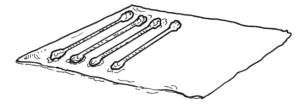

6. Caution: Do not allow children to prepare the cotton swabs. The fumes from the rubber cement can be harmful. Once the swabs set overnight, the fumes disappear.

Procedure
Part One
1. Tell the students that they are going to learn some building techniques that will help them construct some geometric shapes. Distribute the drinking straws and chenille stem pieces.
2. Show students how to bend the chenille stems into right angles. Direct the students to take four straws and to place one chenille stem piece into one end of each straw.

3. Direct them to insert the chenille stem of one straw into the open end of another straw, and continue this process until they form a square.

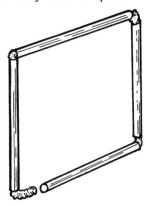

4. Talk about the shape they have just made. Discuss the number of sides, the fact that they are all the same size, etc.

5. Show the students how to make a "table" out of this square by inserting additional chenille stem pieces into the corners of the square.

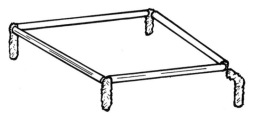

6. Show the students how to insert additional chenille stem pieces into these straws, the legs of the table, and to attach more straws to form a cube.

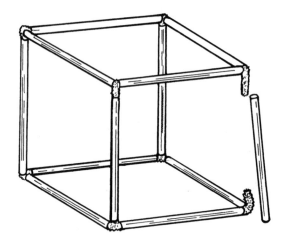

7. Lead the students in a discussion about how many faces, edges, and vertices the cube has.
8. Ask the students to name other objects with this shape.

Part Two
1. Leave additional materials out at a discovery center for the students to add to their cubes. Encourage the students to describe their structures and to name them. For example: a house, a bridge, a tunnel, a space station.
2. Allow the students to build and rebuild and to design and redesign their structures.
3. Display their structures and allow them to either verbally share their processes of building or write instructions as to how or why they used this design.
4. Ask the students to draw their cubes on a sheet of paper. This is a difficult task and will prove to be a challenge to most students.

Part Three
1. Give each student a bag of prepared cotton swabs. Direct them to build a three-dimensional structure using these cotton swabs.

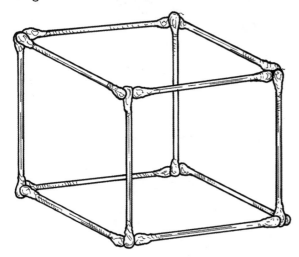

2. Offer other building materials such as Lincoln Logs, LEGO® elements, etc., to further explore the construction of three-dimensional objects.

Connecting Learning
1. How did you build your structure?
2. What could your structure be used for?
3. Is your structure strong enough for you to use? Is it a toy or something for your friends to use too?
4. If you could build this structure out of something else, what would you like to try using?
5. Did you build your structure several times? How did it get better?
6. What did you learn while building your structure?

* Reprinted with permission from *Principles and Standards for School Mathematics*, 2000 by the National Council of Teachers of Mathematics. All rights reserved.

3-D Designs

Key Questions

What kind of three-dimensional structures can we make?

Learning Goals

Students will:

- build a cube using chenille stems and straws, and
- use this figure as a base to explore the building process further.

3-D Designs

Connecting Learning

1. How did you build your structure?

2. What could your structure be used for?

3. Is your structure strong enough for you to use? Is it a toy or something for your friends to use too?

4. If you could build this structure out of something else, what would you like to try using?

5. Did you build your structure several times? How did it get better?

6. What did you learn while building your structure?

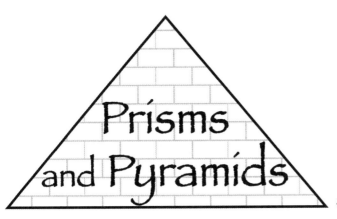

Topic
Pyramids and prisms

Key Question
How are pyramids and prisms alike and how are they different?

Learning Goals
Students will:
- construct models of pyramids and prisms,
- sort them into groups using different rules,
- compare and contrast characteristics of these solids, and
- develop definitions for pyramids and prisms.

Guiding Documents
Project 2061 Benchmarks
- *Assemble, describe, take apart and reassemble constructions using interlocking block, erectors sets, and the like.*
- *Circles, squares, triangles, and other shapes can be found in things in nature and in things that people build.*
- *When trying to build something or get something to work better, it usually helps to follow directions if there are any or to ask someone who has done it before for suggestions.*
- *Several steps are usually involved in making things.*

NCTM Standards 2000*
- *Recognize, name, build, draw, compare, and sort two- and three-dimensional shapes*
- *Classify two- and three-dimensional shapes according to their properties and develop definitions of classes of shapes such as triangles and pyramids*
- *Build and draw geometric objects*

Math
Geometry
  3-D solids
    pyramids, prisms

Integrated Processes
Observing
Comparing and contrasting
Identifying

Materials
Drinking straws (see *Management 2*)
Chenille stems (see *Management 2*)
Unifix cubes (see *Management 2*)
Scissors
Station cards
Student page

Background Information
In 3-D geometry, a prism is a solid that has two congruent, parallel bases, each of which is a polygon. These bases are connected by lateral faces that are parallelograms. The prisms that are generally introduced in the elementary school curriculum are regular, right prisms. These are prisms where the bases are regular polygons and the lateral faces are rectangles that meet the bases at right angles.

All rectangular solids, including cubes are prisms. These are usually called rectangular prisms, even when their bases are square. Other regular, right prisms include the hexagonal and triangular prisms pictured here.

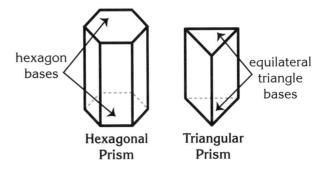

Hexagonal Prism — hexagon bases
Triangular Prism — equilateral triangle bases

Unlike prisms, pyramids have only one base. This base is a polygon, and all of the faces of a pyramid are triangles. These triangular faces meet at a common point (vertex), known as the apex. Elementary students are most commonly exposed to two regular

SHAPES, SOLIDS, AND MORE © 2009 AIMS Education Foundation

pyramids (those with regular polygons for a base)—the square pyramid, and the triangular pyramid. If the triangular pyramid is composed entirely of equilateral triangles, then it is also considered a tetrahedron, but not all triangular pyramids are tetrahedrons.

Square   Triangular   Tetrahedron
Pyramid   Pyramid

Management
1. Students need to work in groups of four. Each group will construct four models, one of each type.
2. Each group needs 14 drinking straws, nine chenille stems, and seven Unifix cubes. The narrower the straw, the more snugly the chenille stems will fit. Be sure to get the standard chenille stems that have about a ¼-inch diameter so that they will fill the straws.
3. Make each model before doing the activity. Construction instructions are found in the *Procedure* and on the station cards. These models will be used by students as samples when they construct their own.
4. This activity is divided into multiple parts. The first three parts involve the construction of the models. The fourth part involves exploration of the models and the development of definitions. It is suggested that you do the parts across two or even three days.

Procedure
Part One: Preparing the Materials
1. Divide students into groups and distribute the materials. Explain that they will be constructing four models of geometric solids within their groups; each student will be responsible for making one of the models, but they will work together to prepare the materials needed.
2. Invite a student to come to the front of the room to help you demonstrate how to cut the straws into equal lengths. As the student is coming up, stack three Unifix cubes together. Show the class how to put the straw through the holes in the center of the Unifix cube tower. Place the tower on a table or desk and point out that the straw goes all the way to the bottom of the tower; it is touching the table.
3. Ask the student who has come to the front of the room to hold the tower of Unifix cubes with one hand and the straw with the other hand. (The straw should be held upright, not tilted at an angle.) Cut the straw even with the top of the Unifix cube tower. Remove the cut piece, set it aside, and place the remaining length of straw in the tower, being sure that it goes all the way to the table. Repeat until the straw has been cut into three equal pieces with a small bit left over.
4. Explain that students will need to work in pairs like this to cut all the straws to specific lengths. One person will hold the straw and the tower, the other will cut the straw. Emphasize that the straw must be all the way through the tower of cubes so that it touches the table.
5. Have the students pair up within their groups and decide which pair will cut the four-cube lengths and which will cut the three-cube lengths. Have each group take seven straws and cut them into as many lengths as they can. (There should be 21 three-cube lengths and 14 four-cube lengths.) Be sure they realize that there will be small pieces of straw left over each time; these pieces should be discarded.

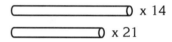

6. When all the straws have been cut, instruct students to repeat the process with the chenille stems, this time, cutting them to two-cube lengths. Each group needs three uncut chenille stems, three chenille stems that have had one two-cube length cut from them, and 20 two-cube length pieces. There will be about half of a chenille stem left over. This can be set aside in case any are lost.

Part Two: Exploring Construction Techniques
1. Explain that students will be using the materials they have prepared to make models of three-dimensional shapes.
2. Instruct two students in each group to take a long (uncut) chenille stem and the remaining two students to take a medium-length chenille stem.
3. Demonstrate how to thread the short straw pieces onto the chenille stem. Have the students with the long stems thread four short straw pieces, and those with the medium-length stems thread three short straw pieces. Have students push the straw pieces close together and center them so that there is extra chenille stem on each end.

4. Show students how to bend the chenille stem to form either a square or a triangle, depending on the number of straws.

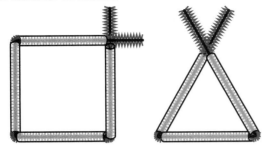

5. Have students bend the ends of the chenille stems that are sticking out from the ends of the straws. Show them how to insert one of the bent ends into the adjacent straw and bend the other one so that it is perpendicular to the square or triangle.

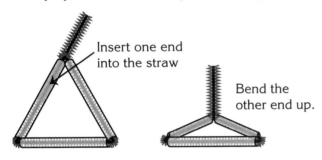

Insert one end into the straw

Bend the other end up.

6. Have each student take a small piece of chenille stem and bend it to form a right angle. Instruct them to place a straw on one end. Explain that this is another way that straws can be connected. Have them put the exposed end of the chenille stem into one of the straws on the triangle or square they made so that the straw is perpendicular to the shape.

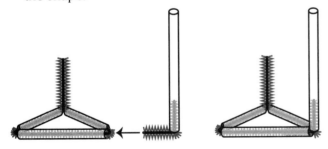

Part Three: Constructing the Models
1. Show groups the four completed models you made. Explain that each student within a group will be responsible for making one of the four models. They will use what they have just learned and the instructions on the station cards to make their own.
2. Set each model on a table along with the corresponding station card. Point out the two models that will make use of the squares students just made and the two that will make use of the triangles. Inform them that they will use the square or triangle they have already made as part of the model, and that remaining materials needed are illustrated on the station cards.
3. Instruct students to decide within their groups who will make what model. (Those who made the squares will need to make the one of models that use a square and those who made triangles will need to make one of the models that uses a triangle.)
4. Have students go to the table with the appropriate station card, determine the materials needed, and collect them from the collection of materials their group prepared in *Part One*.
5. Allow time for the students at the tables to work together to make their models, then return to their groups when they have finished.

Part Four: Exploring Prisms and Pyramids
1. Have students get out their models and trade them with group members so that they can look at the similarities and differences among the models.
2. Tell students to divide the models into two groups. They can use whatever rule they want, but they must be able to explain the rule.
3. Invite a student from one group to come to the front of the class and use the models you made for demonstration to show how his/her group divided the models. Have that group share the rule they used for the grouping. Record the rule used for the grouping on the board. (e.g., some triangular faces/no triangular faces or, comes to a point/doesn't come to a point)
4. Ask if any other groups decided on a different rule to sort the models. Continue to invite students to the front to share until no more ways to divide the models into two groups can be found.
5. Tell the students that mathematicians would divide the models they made into two groups called pyramids and prisms. Ask students to guess which models would be in each group and why they think so.
6. Show the students which two models are prisms and which two are pyramids. Give each group a copy of the student page. Instruct them to study the models and to write down all the ways that the prisms are like each other and all the ways that the pyramids are like each other. Encourage them to describe the faces and to use appropriate geometric vocabulary such as congruent, parallel, perpendicular, edge, vertex, etc.
7. Invite groups to share their lists. Develop class lists of the characteristics of prisms and of pyramids. Add any characteristics that students may not have noticed and correct any erroneous information students may have listed (e.g., all pyramids have either square or triangular bases).

8. As a class, develop definitions of prisms and pyramids using students' observations as a starting place. (See *Background Information*.) Introduce the term *base* as it applies to prisms and pyramids. Be sure students understand why cylinders are not prisms and cones are not pyramids. [Their faces are not polygons.]
9. Challenge students to determine which of the shapes shown on the bottom of the page are prisms, which are pyramids, and which are neither.

Connecting Learning
1. What were some of the different ways that you sorted your models?
2. What are some ways that your models of prisms are alike? [They have two congruent faces that are the bases; the bases are connected by faces that are rectangles; they are the same height, etc.] How are they different? [One has triangles and the other doesn't.]
3. What are some ways that your models of pyramids are alike? [They have faces that meet at a common point; the faces that meet at a point are all congruent triangles; etc.] How are they different? [One has a square base, the other has a triangular base; they are different heights; etc.]
4. How can you tell if a shape is a prism? [Prisms have two faces (the bases) that are congruent polygons. These bases are parallel. The remaining faces are all quadrilaterals.]
5. How can you tell if a shape is a pyramid? [Pyramids have one face (the base) that is a polygon. The remaining faces are all triangles and meet at a common point (the apex).]
6. Why aren't cylinders prisms? [They don't have polygons for bases; they don't have faces that are quadrilaterals.] Why aren't cones pyramids? [They don't have a polygon for a base; they don't have faces that are triangles.]
7. Which shapes on the page were prisms? ...pyramids? ...neither? [The cone, cylinder, and sphere are not prisms or pyramids.]

* Reprinted with permission from *Principles and Standards for School Mathematics*, 2000 by the National Council of Teachers of Mathematics. All rights reserved.

Prisms and Pyramids

Key Question

How are pyramids and prisms alike and how are they different?

Learning Goals

Students will:

- construct models of pyramids and prisms,
- sort them into groups using different rules,
- compare and contrast characteristics of these solids, and
- develop definitions for pyramids and prisms.

Prisms and Pyramids
Model Construction

You need:

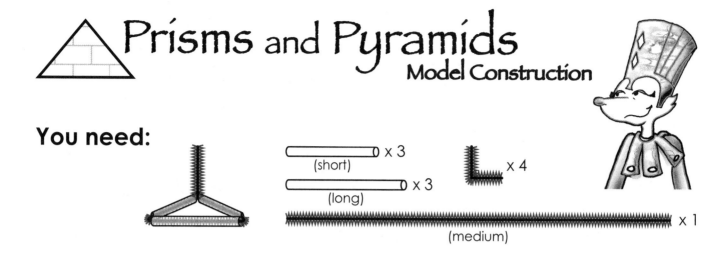

First, make another triangle.

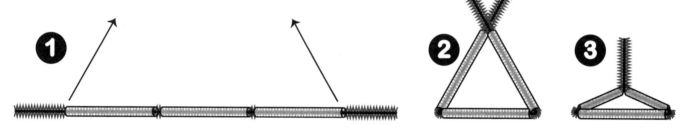

Next, add the connectors to both triangles.

Then, connect the triangles.

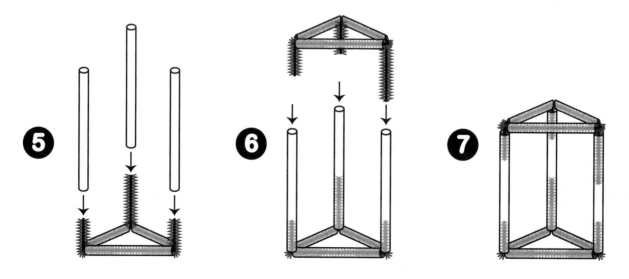

SHAPES, SOLIDS, AND MORE © 2009 AIMS Education Foundation

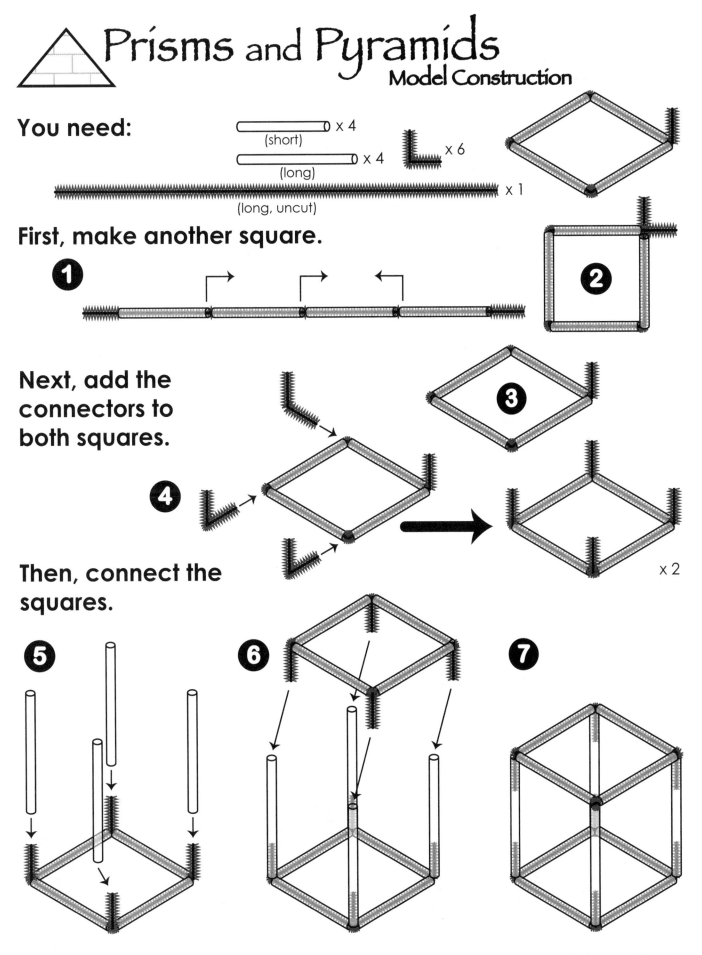

Prisms and Pyramids
Model Construction

You need:

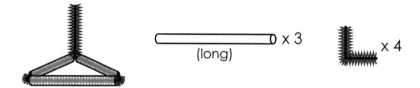

First, add the connectors to the triangle.

1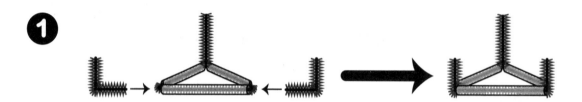

Next, put a straw on each connector.

2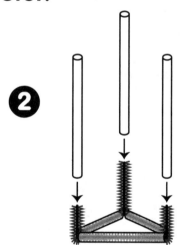

Then, bend the straws to meet in the middle.

3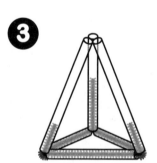

Last, connect pairs of straws with the connectors.

4 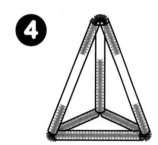   **5**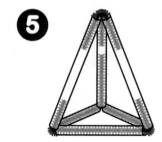

SHAPES, SOLIDS, AND MORE © 2009 AIMS Education Foundation

Prisms and Pyramids
Model Construction

You need: 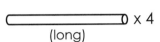 x 4 (long)   x 6

First, add the connectors to the square.

1

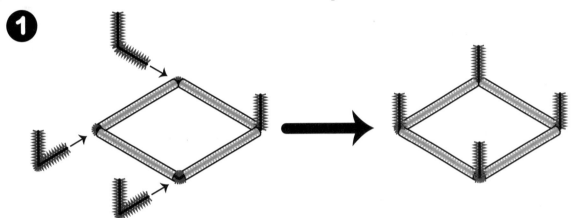

Next, put a straw on each connector.

2

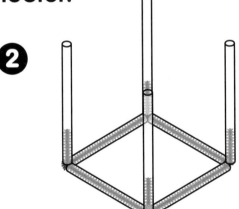

Then, bend the straws to meet in the middle.

3

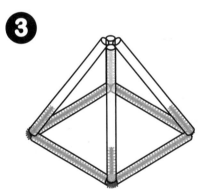

Last, connect pairs of straws with the connectors.

4 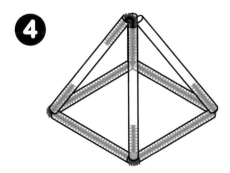   **5**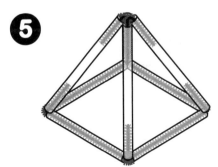

SHAPES, SOLIDS, AND MORE   333   © 2009 AIMS Education Foundation

Prisms and Pyramids

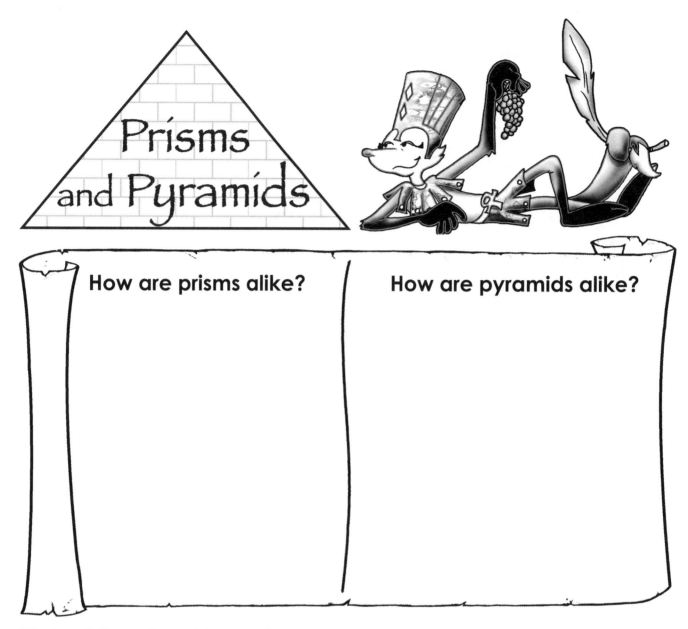

How are prisms alike?

How are pyramids alike?

Circle all the prisms. Make a box around all the pyramids. Cross out the shapes that aren't prisms or pyramids.

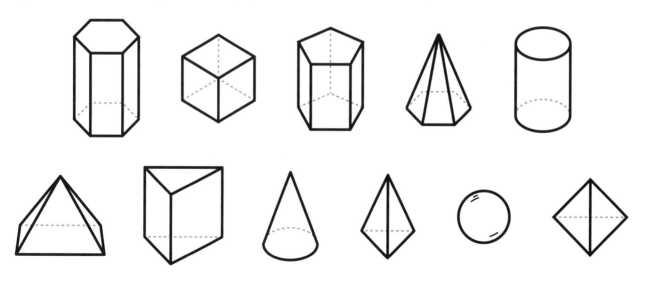

SHAPES, SOLIDS, AND MORE

Connecting Learning

1. What were some of the different ways that you sorted your models?

2. What are some ways that your models of prisms are alike? How are they different?

3. What are some ways that your models of pyramids are alike? How are they different?

4. How can you tell if a shape is a prism?

5. How can you tell if a shape is a pyramid?

6. Why aren't cylinders prisms? Why aren't cones pyramids?

7. Which shapes on the page were prisms? …pyramids? …neither?

Get a Clue

Topic
Geometric solids and 2-D shapes

Key Question
How can you locate the object being described using only your sense of touch?

Learning Goal
Students will identify shapes and solids using their sense of touch and verbal clues given by their partners.

Guiding Documents
Project 2061 Benchmarks
- *Numbers and shapes can be used to tell about things.*
- *Use numerical data in describing and comparing objects and events.*

NCTM Standards 2000*
- *Recognize, name, build, draw, compare, and sort two- and three-dimensional shapes*
- *Describe attributes and parts of two- and three-dimensional shapes*
- *Identify, compare, and analyze attributes of two- and three-dimensional shapes and develop vocabulary to describe the attributes*

Math
Geometry
    2-D shapes
    3-D solids

Integrated Processes
Observing
Communicating
Comparing and contrasting

Materials
Pattern blocks
Tangram pieces
3-D solids
Mystery boxes, one per pair of students
List of objects found in each box

Background information
Young children need multiple opportunities to use the vocabulary of geometry in order for them to internalize the terms. This activity provides a playful opportunity for students to explore the properties of both two- and three-dimensional shapes. During the activity, students will be expected to use the correct geometry terms to describe specific shapes and solids for their partners to identify as well as use their sense of touch to identify properties of specific shapes and solids.

Management
1. To make the mystery boxes, choose a cardboard box that is a minimum of eight inches wide, 12 inches long, and six inches high. Cut a five-inch square out of one end of the box so that students can reach in without seeing inside the box. Shoeboxes work well.
2. Gather one mystery box and set of objects for each pair of students prior to doing this activity. Fill each box with a variety of two- and three-dimensional objects such as pattern block shapes (item number 4250), tangram pieces (item number 4180), and Geo-Solids (item number 4610). For each mystery box, create a list of the items that are in the box.

Procedure
1. Divide the students into pairs. Give each pair of students a mystery box and a list of objects that are in the box.
2. Explain that one student will have the list of the objects in the box and that he/she will describe a shape or solid found in the box. It will be the partner's job to use the sense of touch to locate the object described. For example, "This object has six square faces, eight corners or vertices, and 12 edges." Encourage students to use the properties they have been learning, such as whether the shape is two or three dimensional; if the object stacks, slides, or rolls; the number of points, edges, faces, and types of faces; whether the shape or solid is symmetrical, etc. Instruct the students to switch roles once the object described has been found.

SHAPES, SOLIDS, AND MORE   © 2009 AIMS Education Foundation

3. Allow time for each partner to have several opportunities to both describe and find objects.
4. End with a discussion about the properties of the shapes and solids students were able to identify using only their sense of touch.

Connecting Learning
1. Was it easier to identify two- or three-dimensional shapes? Why?
2. Was it easy to determine if the shape was symmetrical? How did you decide?
3. What real-world objects could we use to replace the sphere? ...the cube? ...cone? ...cylinder? ...square? ...circle?
4. What clues gave you the most information?
5. What solid has a square face and four triangular faces? [square-based pyramid]
6. When might you try to locate something you can't see by its shape? [a pencil in the bottom of your backpack, a toy on the bottom of the toy box, a diving ring, etc.]

* Reprinted with permission from *Principles and Standards for School Mathematics,* 2000 by the National Council of Teachers of Mathematics. All rights reserved.

Get a Clue

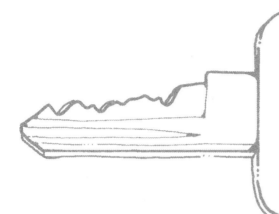

Key Question

How can you locate the object being described using only your sense of touch?

Learning Goal

Students will:

identify shapes and solids using their sense of touch and verbal clues given by their partners.

Get a Clue

Connecting Learning

1. Was it easier to identify two- or three-dimensional shapes? Why?

2. Was it easy to determine if the shape was symmetrical? How did you decide?

3. What real-world objects could we use to replace the sphere? ...the cube? ...cone? ...cylinder? ...square? ...circle?

4. What clues gave you the most information?

5. What solid has a square face and four triangular faces?

6. When might you try to locate something you can't see by its shape?

Topic
Lines and locations

Key Question
How can we locate specific geometric objects on a map?

Learning Goals
Students will:
- find and name locations of shapes, with simple relationships such as near to, beside, in front of, etc.;
- identify shapes by their names and attributes; and
- describe paths using directional words such as left, right, forward.

Guiding Documents
Project 2061 Benchmark
- *Raise questions about the world around them and be willing to seek answers to some of them by making careful observations and trying things out.*

NRC Standard
- *The position of an object can be described by locating it relative to another object or the background.*

NCTM Standards 2000*
- *Describe attributes and parts of two- and three-dimensional shapes*
- *Recognize, name, build, draw, compare, and sort two- and three-dimensional shapes*
- *Recognize and represent shapes from different perspectives*
- *Organize and consolidate their mathematical thinking through communication*
- *Describe location and movement using common language and geometric vocabulary*

Math
Geometry
Problem solving
Number sense

Integrated Processes
Observing
Comparing and contrasting
Communicating
Relating
Collecting and recording data

Materials
For each group:
  floor map (see *Management 2*)
  small toy vehicles (see *Management 3*)
  pattern blocks

For the class:
  Shape Town story

For each student:
  Shape Town Tally Sheet
  Shape Town Map

Background Information
Mathematical questions regarding navigation and maps can help students develop a variety of spatial understandings: direction, distance, location, and representation. Students develop the ability to navigate first by noticing landmarks, then by building knowledge of a route, and finally by putting many routes and locations into a kind of mental map.* (NCTM, p.98)

The map setting in this activity will give the students an opportunity to develop navigation skills as they relate to the real world. It also gives an opportunity for the students to review and reinforce geometric terminology such as parallel, intersecting, straight, and curved lines as well as the names of several two-dimensional shapes.

Using small vehicles to roam the streets of the map allows children to look at the map from the vehicle's perspective and not from the more difficult bird's eye view.

While parallel and intersecting lines are not the entire focus of this activity, the map setting is a way to connect students' experiences with streets to these two abstract terms. Students will have experienced driving through the intersections of two roads.

Management
1. This activity is designed for a small group setting.
2. Enlarge the map to poster size or purchase an AIMS Floor Map at aimsedu.org (item number 1520S). Laminate the floor map for extended use.
3. You will need to gather small toy vehicles to use on the map. Small cars, finger scooters, and finger skateboards can be purchased at your local department store.
4. Make copies of the recording sheet and review the tally process.

Procedure

1. Invite a small group of students to an area where you can spread out.
2. Read the story *Shape Town*.
3. Tell the students in the group that they are going to create a town where the shapes live happily together. Ask your students to use pattern blocks to create buildings that combine shapes and place them on the map. Invite the students to name the buildings and the streets. Use a wax pencil or overhead marker to write these names directly on the streets and beside the buildings.

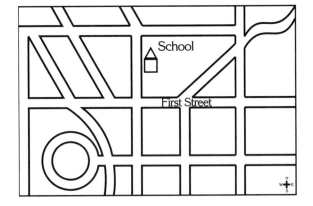

4. Ask the *Key Question*.
5. Have one student put his or her vehicle on the map. Ask what path would need to be taken to go from one location to another. Question the student about the shapes that are passed, the direction the vehicle is going, whether the path will be straight or curved, and if any types of turns will be made.
6. Direct another student to put his or her vehicle on the map, and ask the other students to give directions to get from one point to another.
7. Find two streets that intersect and tell one student to put a vehicle on one of the streets and another student to put his or her vehicle on the other street. Ask the students if it would be possible for the two people to play together without leaving their streets.
8. After all of the students have had an opportunity to travel on the map and describe their path, give each student an assigned starting point and an ending point. Using the *Shape Town Tally Sheet*, ask them to use tallies to record the shapes they pass along the way.
9. Compare the tally sheets.

Connecting Learning

1. Can you get from the _____ to the _____ without making any left turns?
2. How did you decide which path to take?
3. Was the shape of your path curved, straight, or a combination of the two?
4. What shapes were found on our map?
5. What shapes did you pass most often?
6. To get from _____ to _____, I want you to make a left turn. What path could you take? Are there other paths you could take?
7. What shapes can be found at the corner of _____ and _____?
8. If I travel along _____ and you travel along _____, will our paths cross? (Name a pair of intersecting streets so the answer will be yes.)
9. Describe two paths that would be parallel.
10. Describe the path you might take to get from ____ to ____ that is both straight and curved.
11. Go from _____ to _____ and describe your path.
12. (Clear the pattern blocks off.) If you look closely, the roads form different shapes. Where can you find a triangle on the map? ...trapezoid? ...parallelogram? ...circle? ...square? ...rectangle?

Extensions

1. Prepare a floor map of other settings like an amusement park, etc. It should be large enough that a small toy vehicle can easily maneuver on the roads. It should also have empty areas for the children to place the shapes and street labels. Street labels can be written on 3" x 5" cards and glued to craft sticks that are mounted into small balls of clay to give the map a three-dimensional feeling.
2. Ask the students to create their own maps and questions.

Curriculum Correlation

Seuss, Dr. *The Shape of Me and Other Stuff.* Random House. New York. 1973.
Rhyme and silhouette drawings introduce the shape of bugs, balloons, peanuts, camels, spider webs, and many other familiar objects.

* Reprinted with permission from *Principles and Standards for School Mathematics,* 2000 by the National Council of Teachers of Mathematics. All rights reserved.

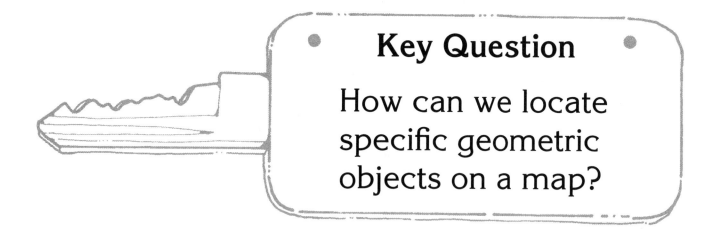

Key Question

How can we locate specific geometric objects on a map?

Learning Goals

Students will:

- find and name locations of shapes, with simple relationships such as near to, beside, in front of, etc.;
- identify shapes by their name and attributes; and
- describe paths using directional words such as left, right, forward.

Once upon a time, in a place called Shape Land, there lived a Shape Seeker. The Shape Seeker traveled all over the land looking for shapes. The Shape Seeker would travel straight paths, curved paths, parallel paths, and intersecting paths.

The Shape Seeker discovered that there were different towns in the land and that in each town lived different shapes. There was Triangle Town, Circle Town, Square Town, and Rectangle Town, to name just a few. Most of the time the shapes in the town were happy and playful, but sometimes they were frustrated because they could not build some of the things they needed. You see, there were only triangles in Triangle Town and only squares in Square Town and so on. There were no wagons in Circle Town because there were no rectangles! There were no bicycles in Square Town because there were no circles! Oh my! The Shape Seeker had an idea. Why not start a new town, one called Shape Town?

Invitations were sent out to all the towns to join the Shape Seeker at a picnic in the new town of Shape Town. On the day of the picnic, some shapes traveled along straight paths to get to Shape Town. Others traveled on curved paths. Some had more than one path; they traveled on parallel paths, while others took intersecting paths.

When they all arrived, it was quite a sight! Squares had never seen circles before, and rectangles had never seen triangles. Triangles had never seen squares, and circles had never seen rectangles! All the shapes just stood and stared!

The Shape Seeker knew something must be done to get the shapes to discover the wonderful things they could do if they would work together. Music began to play, and the shapes began to dance. They twirled and turned, they jumped and hopped, and they began dancing all around Shape Town. They danced and danced until they were exhausted. All of a sudden, the shapes grew very quiet. They heard a strange sound. The rectangle was groaning as it fainted and started to fall. Four circles that were nearby rolled over to catch the rectangle. Everyone looked and then began to cheer. A wagon was made when the rectangle fell on top of the circles! The rectangle came to and saw what had happened. It was so grateful to be held up by the circles. Two triangles jumped onto the wagon and they rolled all around Shape Town in celebration.

The rest of the shapes became excited and wanted to make something. They joined together and made a train, a bridge, a house, and many more wonderful things.

The Shape Seeker watched as the shapes from around the land began to work together and build things for their new town called Shape Town.

Tally Sheet

What shapes did you pass on your journey?

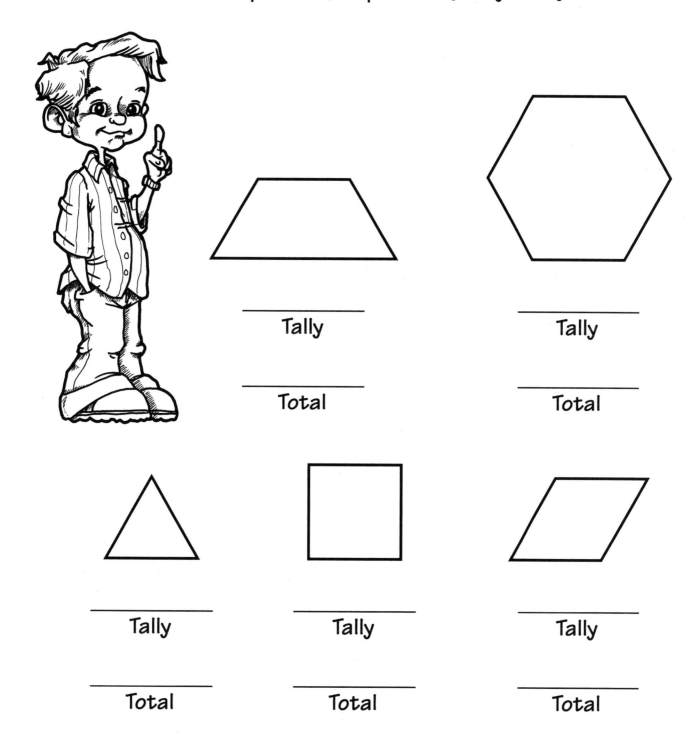

SHAPES, SOLIDS, AND MORE

© 2009 AIMS Education Foundation

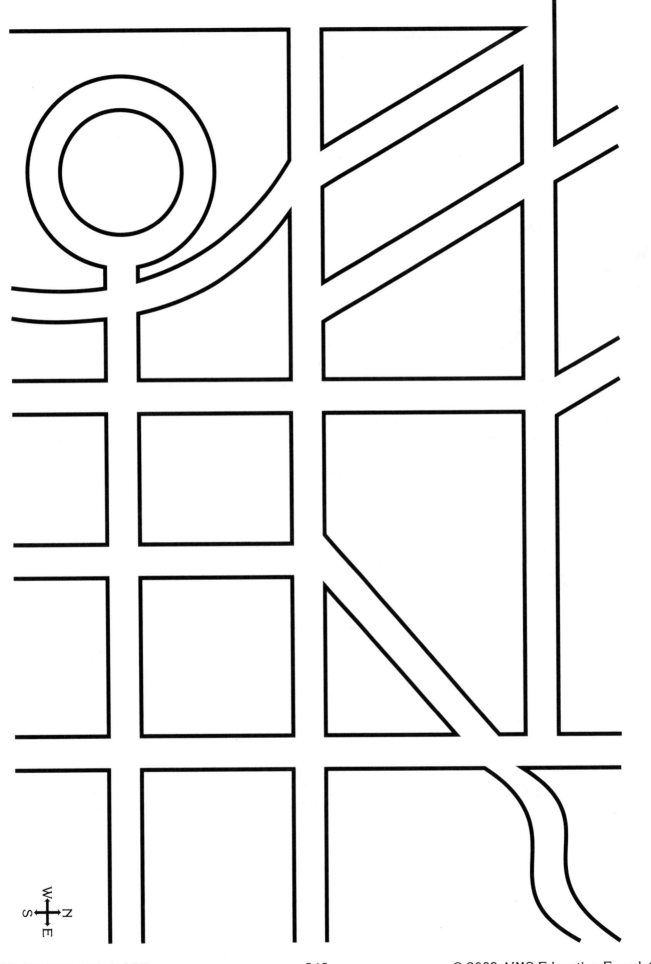

SHAPES, SOLIDS, AND MORE   346   © 2009 AIMS Education Foundation

Connecting Learning

1. Can you get from the _____ to the ____ without making any left turns?

2. How did you decide which path to take?

3. Was the shape of your path curved, straight, or a combination of the two?

4. What shapes were found on our map?

5. What shapes did you pass most often?

6. To get from _____ to _____, I want you to make a left turn. What path could you take? Are there other paths you could take?

7. What shapes can be found at the corner of _____ and _____?

Connecting Learning

8. If I travel along _____ and you travel along _____, will our paths cross?

9. Describe two paths that would be parallel.

10. Describe the path you might take to get from ____ to ____ that is both straight and curved.

11. Go from ____ to ____ and describe your path.

12. If you look closely, the roads form different shapes. Where can you find a triangle on the map? …trapezoid? …parallelogram? …circle? …square? …rectangle?

Where Is It?

Topic
Positional words

Key Question
What positional words can we use to describe an object's location?

Learning Goals
Students will:
- place objects on shelves based on descriptions of their positions relative to other objects, and
- describe the positions of objects relative to other objects.

Guiding Documents
NRC Standard
- The position of an object can be described by locating it relative to another object or the background.

*NCTM Standards 2000**
- *Describe, name, and interpret relative positions in space and apply ideas about relative position*
- *Describe, name, and interpret direction and distance in navigating space and apply ideas about direction and distance*
- *Find and name locations with simple relationships such as "near to" and in coordinate systems such as maps*
- *Describe location and movement using common language and geometric vocabulary*

Math
Geometry
   position

Integrated Processes
Observing
Comparing and contrasting
Communicating
Applying

Materials
Transparency (see *Management 1*)
Objects to manipulate (see *Management 2*)
Scissors
Glue
Student pages

Background Information
Positional words and phrases, such as *on, above, below, between, beside, to the left of, to the right of,* etc., are used to describe where things are located. Some positional words are more specific than others. For example, if I tell you that my keys are in my bedroom, you are less likely to find them than if I tell you that they are on my nightstand. This activity is designed to provide students with an opportunity to practice applying positional words as they describe the locations of specific objects as well as locate objects based on descriptions that use positional words. It is also a good exercise in listening and following directions.

Management
1. Make an overhead transparency of the page of grocery items on the store shelves.
2. Prior to teaching this lesson, arrange several items on a desk or table where the entire class can see. It is suggested that you stack various books with the labels facing the students and place items like a stapler, coffee cup, tape dispenser on either side of the books.

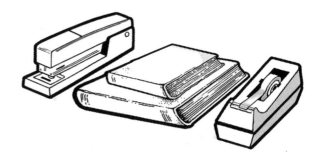

3. Directional words such as north, south, east, west can be used in this activity if you choose.

Procedure
1. Ask student what words they use when describing the position of something. Record the list on the board. Ask how they might describe where a classroom item, such as the pencil sharpener, is located. Explain that words like under, over, beside, between, etc., are called positional words because they describe the position of an object.
2. Tell the class that you would like to see how good they are at using positional words.

SHAPES, SOLIDS, AND MORE

3. Draw students' attention to the display of items prepared ahead of time. Have them describe the location of specific items on display. Look for the correct use of positional words. Discuss how some positional words are more helpful in locating an item than others. Give them an example of less helpful descriptions such as "the stapler is beside the stack of books." Explain that its position might be better described as being on the right or left side of the stack of books. Allow several students to describe the positions of various items.

4. Invite students to rearrange the items on display based on specific directions (e.g., place the spelling book under the math book, place the pencil sharpener on the math book, etc.).

5. Display the transparency referred to in *Management* on the overhead. Focus students' attention on the items on the store shelves. Ask questions such as: What is directly below the Coco Crunch? What is to the left of the jam? Etc.

6. After each student has an opportunity to identify objects based on your descriptions, identify a specific item and invite students to use positional words to describe its location.

7. When students are comfortable using positional words, distribute the *Part Two* student pages. Ask them to cut out the items on the bottom of the first page. Explain that you would like them to stock the store shelves based on the directions given on the second page. Allow time for students to work.

8. When students have completed stocking the shelves, have them compare their results with other students.

9. End with a discussion about the use of positional words and why it is important to be as specific as possible.

Connecting Learning
1. What type of words do we use to describe the position of something?
2. How would you describe the position of my teacher's desk?
3. Why is it important to be very specific when describing the position of something?
4. If I told you that my whistle was somewhere in the classroom, would it be easy for someone to find it? Why, or why not?

Extension
Play a version of "I Spy" that involves location clues instead of color clues.

Solution

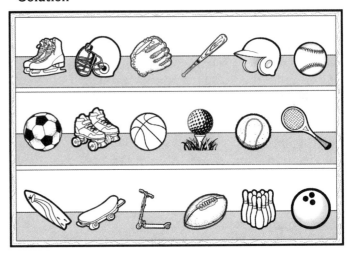

* Reprinted with permission from *Principles and Standards for School Mathematics,* 2000 by the National Council of Teachers of Mathematics. All rights reserved.

Where Is It?

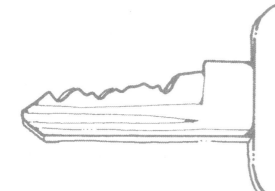

Key Question

What positional words can we use to describe an object's location?

Learning Goals

Students will:

- place objects on shelves based on descriptions of their positions relative to other objects, and
- describe the positions of objects relative to other objects.

Where Is It?

Where Is It?  Part Two

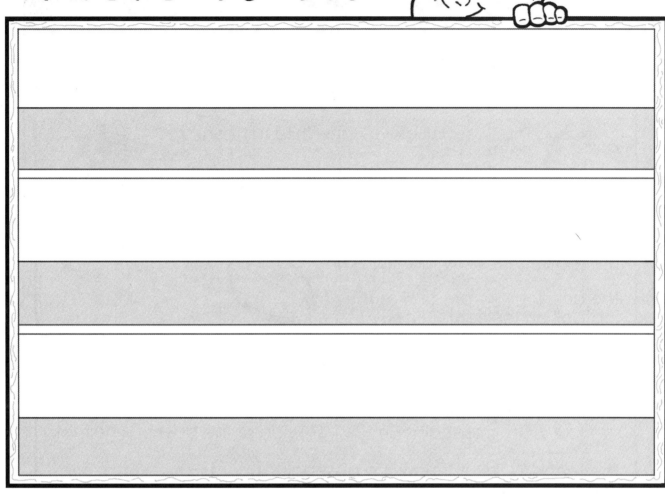

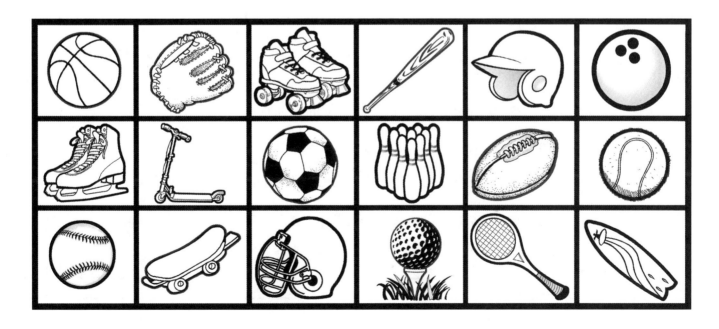

SHAPES, SOLIDS, AND MORE © 2009 AIMS Education Foundation

Where Is It?  Part Two

The ice skates are at the far left on the top shelf.

 The baseball glove is the third item on the top shelf.

The bat is to the right of the ball glove.

The golf ball is on the shelf below the baseball bat.

The football is below the golf ball.

The bowling pins are beside the football on the right.

The tennis ball is above the bowling pins.

 The baseball helmet is above the tennis ball.

The baseball is on the right side of the baseball helmet.

The tennis racket is below the baseball.

The bowling ball is below the tennis racket.

The basketball is below the baseball glove.

The football helmet is between the ice skates and baseball glove.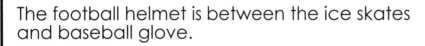

The roller skates are below the football helmet.

The surfboard is in the far left corner of the bottom shelf.

The skateboard is between the surfboard and the scooter.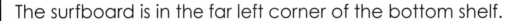

 The soccer ball is above the surfboard.

The scooter is beside the football.

Where Is It?

Connecting Learning

1. What type of words do we use to describe the position of something?

2. How would you describe the position of my teacher's desk?

3. Why is it important to be very specific when describing the position of something?

4. If I told you that my whistle was somewhere in the classroom, would it be easy for someone to find it? Why, or why not?

Coordinate Camping

Topic
Coordinate grid

Key Question
How can you use a coordinate grid to locate things?

Learning Goals
Students will:
- locate camping objects on a coordinate grid, and
- identify areas in camp that are at specific coordinate points.

Guiding Document
NCTM Standards 2000*
- *Make and use coordinate systems to specify locations and to describe paths*
- *Find the distance between points along horizontal and vertical lines of a coordinate system*
- *Describe, name, and interpret relative positions in space and apply ideas about relative position*
- *Find and name locations with simple relationships such as "near to" and in coordinate systems such as maps*

Math
Geometry
   coordinate grid

Integrated Processes
Observing
Applying

Materials
The Fly on the Ceiling (see *Curriculum Correlation*)
Overhead transparencies (see *Management 1*)
Student pages
Various small objects (see *Procedure, Part One, 2*)

Background Information
   Our world is full of opportunities to use a Cartesian coordinate system. Treasure hunters use grids to search the bottom of the ocean, astronomers use a grid system to map the skies, even entire cities, like Chicago, are built on a grid system.
   The Cartesian coordinate system consists of two number lines set perpendicular to each other. The horizontal axis is called the *x*-axis, and the vertical axis is called the *y*-axis. This allows us to represent any location on a coordinate by a point. The point's location consists of a pair of numbers; the first is the *x*-coordinate and the second is the *y*-coordinate. Points are written in the form *(x, y)*. For example, the point (2, 3) is located two units to the right of (0, 0) and three units up.
   This activity was written to familiarize students with the Cartesian coordinate system and to give them practice using the grid in a playful way. They will locate items by giving the coordinates at which they are located and will identify locations on a map when given their coordinates.

Management
1. Make a transparency of the blank coordinate grid and the grid with camping supplies.
2. *Part Two* of this activity can be done independently, in partners, or as an entire class lesson. Each person, group, or class will need a copy of the coordinate camping grid and camp map.
3. If the class works together on *Part Two* of this activity, it is not be necessary to copy the last student page.

Procedure
Part One
1. Read *The Fly on the Ceiling* by Ruth Glass. Discuss the Cartesian coordinate system.
2. Place the coordinate grid transparency onto the overhead. Position various small objects such as a penny, paper clip, etc., at different coordinate points on the grid. Invite students to give you ordered pairs of numbers that give the specific locations of the objects. Move the objects to allow several students the opportunity to identify locations.
3. When students are comfortable using the coordinate system, display the transparency labeled *Camping Supplies*. Discuss the fact that many, but not all, of the items on the page would be necessary on a camping trip.
4. Explain that the class will be divided into two teams and that one player from each team will be asked to identify the location of something that would be used on a camping trip. Tell the class that their team will get one point for each item correctly located and that points will be deducted for locating items that would not be used for camping. Play continues until all objects have been located.

SHAPES, SOLIDS, AND MORE   357   © 2009 AIMS Education Foundation

5. Finish the lesson with a time of sharing what they know about using a coordinate grid.

Part Two
1. Review how to locate things using a coordinate system.
2. Distribute the camp map and the question page. Discuss similarities and differences between the map and the supply sheet.
3. Direct students to answer the questions on the student page.
4. Close with a discussion about when in the real world a coordinate grid might be used.

Connecting Learning
1. How do you locate a specific point on a grid? [first go over, then go up]
2. Where is Lookout Mountain located? [(5, 5)]
3. What is located at (3, 2)? [campfire]
4. When might a coordinate grid be used in the real world?

Extension
The transparent grid can often be placed over a piece of cloth or gift-wrap for additional practice locating objects using ordered pairs.

Curriculum Correlation
Literature
Glass, Ruth. *The Fly on the Ceiling*. Random House. New York. 1998.

* Reprinted with permission from *Principles and Standards for School Mathematics,* 2000 by the National Council of Teachers of Mathematics. All rights reserved.

Coordinate Camping

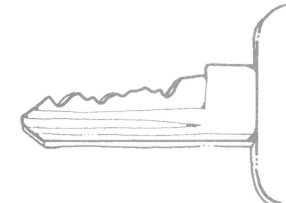

Key Question

How can you use a coordinate grid to locate things?

Learning Goals

Students will:

- locate camping objects on a coordinate grid, and
- identify areas in camp that are at specific coordinate points.

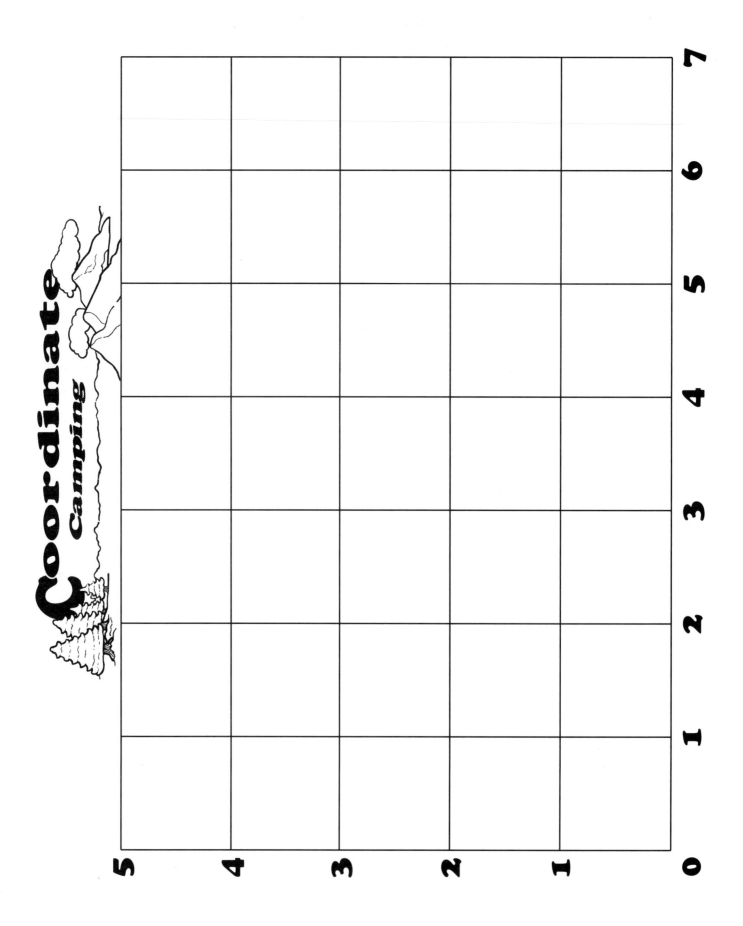

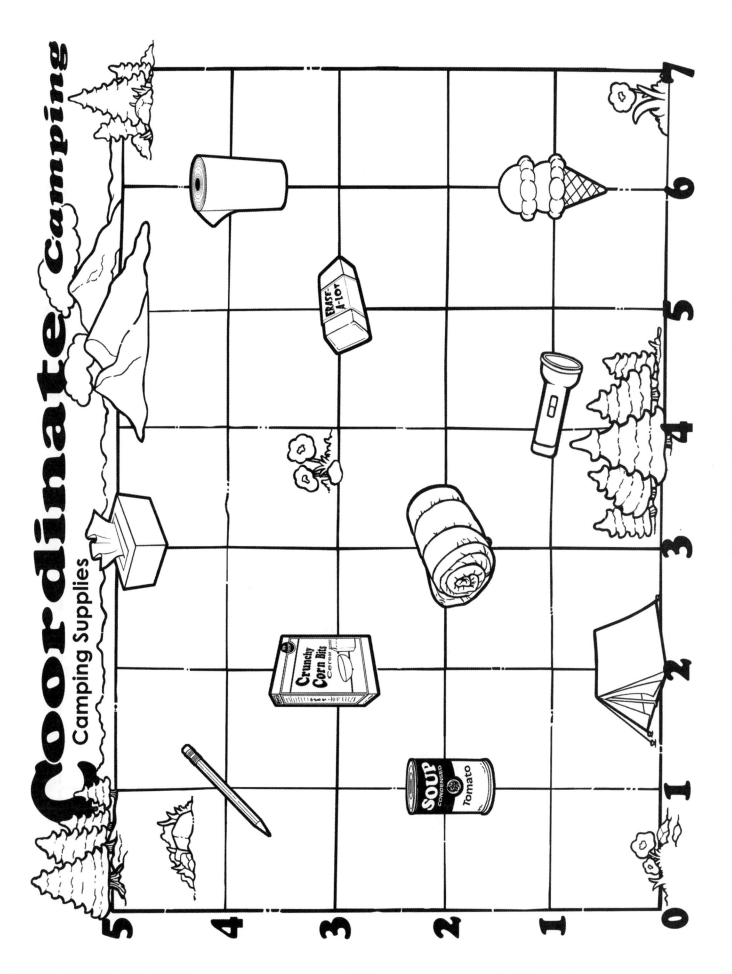

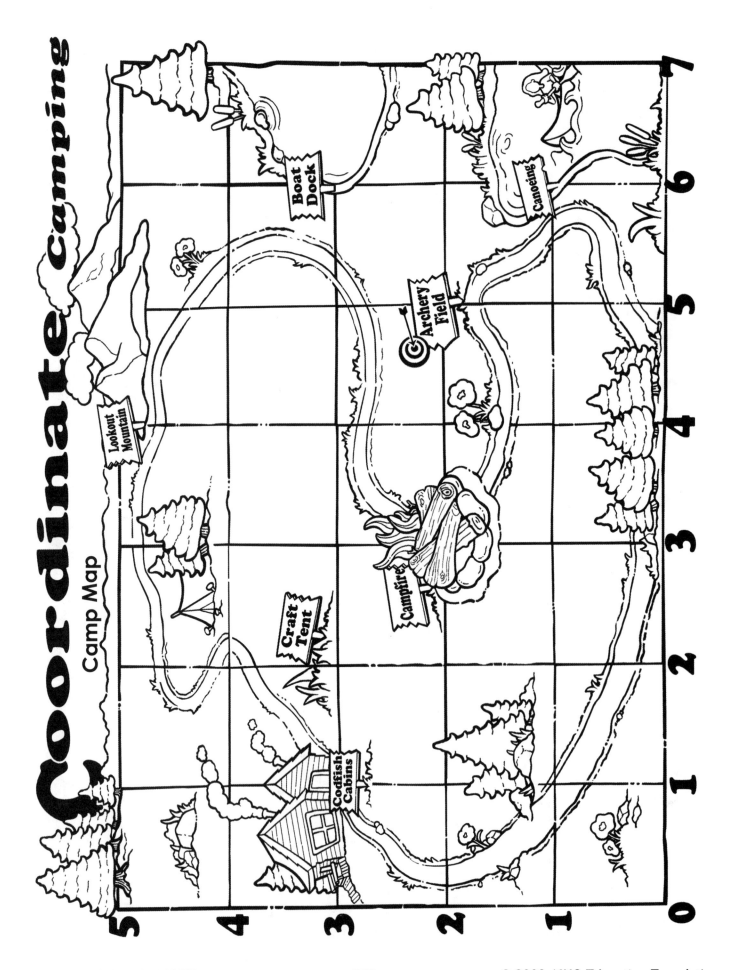

1. If I go to the coordinates (5,2), I will be _____.

2. To go to the boat dock, I would go to (____,____).

3. Where is the campfire located? (____,____)

4. If I am at the cabins, what are my coordinates? (____,____)

5. At what coordinates would I have the best view of the area?

   (____,____)

6. Is the craft tent located at coordinates (3,2) or (2,3)?

SHAPES, SOLIDS, AND MORE   363   © 2009 AIMS Education Foundation

Coordinate Camping

Connecting Learning

1. How do you locate a specific point on a grid?

2. Where is Lookout Mountain located?

3. What is located at (3, 2)?

4. When might a coordinate grid be used in the real world?

Coordinate Chorus

Words by Suzy Gazlay and Kathy Jo Graley   Tune: Frère Jacques

First go o-ver; first go o-ver; then go up;

then go up. You'll find the co- or-di- nates;

You'll find the co- or-di- nates; o- ver, up; o- ver, up.

Make a new point; make a new point.
Add some more; add some more.
What's the shape we're building; what's the shape we're building;
On the floor; on the floor?

This has three sides; this has three sides.
This has four; this has four.
Corners wide or narrow; corners wide or narrow?
Let's make more; let's make more.

Where is _____; where is _____?
Where is (s)he; where is (s)he?
Tell us her (his) coordinates; tell us her (his) coordinates.
Where is (s)he; where is (s)he?

Ship Shape

Topic
Coordinate grid

Key Question
How can you tell what the shape of the hidden ship is?

Learning Goals
Students will:
- locate the position of shapes on a coordinate grid,
- apply their knowledge of attributes of various shapes by locating shapes on a coordinate grid, and
- use problem-solving strategies to locate shapes on a coordinate grid.

Guiding Documents
Project 2061 Benchmark
- Numbers and shapes can be used to tell about things.

*NCTM Standards 2000**
- *Describe, name, and interpret relative positions in space and apply ideas about relative position*
- *Find and name locations with simple relationships such as "near to" and in coordinate systems such as maps*
- *Recognize and represent shapes from different perspectives*
- *Describe attributes and parts of two- and three-dimensional shapes*
- *Create mental images of geometric shapes using spatial memory and spatial visualization*
- *Build new mathematical knowledge through problem solving*
- *Solve problems that arise in mathematics and in other contexts*
- *Apply and adapt a variety of appropriate strategies to solve problems*
- *Organize and consolidate their mathematical thinking through communication*

Math
Geometry
   2-D shapes
   location
   coordinate grid
Problem solving

Integrated Processes
Observing
Recording data
Communicating
Applying

Materials
For the class:
   floor grid (see *Management 2*)
   Chinese jump ropes (see *Management 3*)
   set of polygons (see *Management 4*)
   transparencies of grid

For each student:
   paper grid
   paper shapes
   crayons

Background Information
    In this lesson, students apply the use of a coordinate system to locate shapes as well as to help define attributes such as number of corners, sides, etc., of geometric shapes. This lesson is an application activity for the coordinate grid. It is assumed the children already know how to locate points on a coordinate grid and how to name them by calling out the ordered pairs, such as (2,3), to define a location on the grid. The labels on each axis are placed next to horizontal and vertical lines to identify points on the grid. In other situations, the labels can be placed in the spaces of the grid to identify areas on the grid. Such label placement is typically used on a map.
    The lesson begins by having students stand on coordinate points and stretch Chinese jump ropes into various geometric shapes. This allows them to review the use of ordered pairs to name points on a grid. The transition to *Part Two*, ship finding on a paper grid, then becomes much easier.
    Part Two of this activity asks the students to set up a problem for a partner to solve. Each student places a geometric shape on a coordinate grid. This shape is described as a ship lost at sea. Their partner's job is to solve the problem of locating the lost ship by calling out coordinates. Problem-solving strategies are needed when the students must try to determine the locations using the least number of guesses of points. Does the child call out random points? Does the child systematically call out points to determine the location of the lost ship? Spatial awareness comes into play as children try to visualize the possible shapes of the lost ships.

SHAPES, SOLIDS, AND MORE   © 2009 AIMS Education Foundation

Management
1. It is necessary that the students already know how to read the axes on a coordinate grid. It is important to note that a *point* on the coordinate grid is the intersection of two lines, one vertical and the other horizontal. Typically, the horizontal axis (*x* axis) is read first, followed by the vertical (*y* axis).
2. Prepare a floor coordinate grid large enough for students to stand at the intersections. Use numbers 0-5 to label the horizontal axis, and the numbers 0-5 to label the vertical axis.

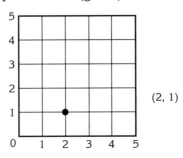

3. Chinese jump ropes can be purchased at most toy stores. If not, a 10-foot length of elastic will work instead of the Chinese jump ropes. Join both ends of the elastic to form a large loop.
4. For step 8 in *Part Two*, duplicate and cut out the set of polygons for the class. Put the polygons in a box or bag from which students will draw. Easier polygons are found on the first page; more difficult ones are on the second page.

Procedure
Part One
1. Gather students around the floor grid. Review how to read a coordinate grid by placing an object on a point on the grid and asking the students to name its location as in point (2, 1) or point (4, 4), etc. Remind them to read the horizontal numbers first, followed by the vertical numbers when naming the points—over then up.
2. Explain that the students are going to use this grid to construct shapes they have been learning about. Direct three students to go to the grid at the following locations, (2, 2), (4, 4), and (5, 2). Explain that these children are the corners of the shape they will be constructing on the grid. Ask the class what shape they think will be made if they connect these points with straight lines. Discuss their responses.
3. Give the students standing on the grid a Chinese jump rope and have them form a corner at the point where each is standing. Discuss how the rope forms the corners and the straight lines. Check what shape is formed and compare with the predictions.

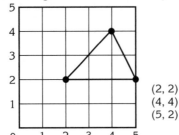

4. Ask the three students to move to three different points on the grid. Call on different students to name the points at which each student is standing, for example, (5, 2), (5, 4), and (3, 3). Ask the class to predict the shape that will be formed when these points are connected with straight lines. Have the students form the shape using the jump rope.
5. Repeat steps 3 and 4 several times, forming different triangles each time. If the students choose three points all on the same line, discuss the difference in this shape compared to the triangles previously formed.
6. Discuss the differences in the corners between the triangles formed. [some are fat/wide, others are skinny/narrow] Discuss the similarities and differences in the triangles. [They all have three corners. They all have three sides. Some have a long side and two shorter sides, some have sides all the same length, etc.]
7. Have the students build another triangle on the grid. Challenge them to determine what they need to do to the jump rope to form it into a rectangle. Allow the students time to solve this problem. Have them explain what they did.
8. Ask students to name the points at which each student is standing on the grid to form the rectangle. Have them describe the locations of the corners, the number of corners, etc.
9. Challenge the students to build many different shapes on the grid including squares, rectangles, parallelograms, trapezoids, etc. Discuss the number of corners, compare the length of sides within each shape, etc.

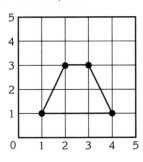

Part Two
1. Give each student a *Ship Shape* grid. Explain that each one needs to secretly draw a three- or four-sided geometric shape on this grid. Tell them to position their shapes so that all corners are on points indicated by two intersecting lines on the grid. Demonstrate placements that would be appropriate and those that would not.

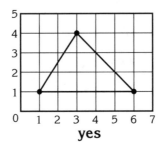
yes

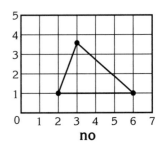
no

SHAPES, SOLIDS, AND MORE 368 © 2009 AIMS Education Foundation

2. Explain that they are going to play a game of *Ship Search*. Tell them to pretend that the shape they have just illustrated is a lost ship in the ocean. Explain that they will sometimes be the captains of rescue teams in search of lost ships and at other times the captains of lost ships.

3. Divide students into groups of two. Tell them to sit across from each other with a divider between them that will hide one grid from another. Designate one partner as the lost ship captain and the other as the rescue captain. Tell the lost ship captains to place the grid they illustrated earlier directly behind the divider, out of their partner's view. Give the rescue captain a blank grid and crayon.

4. Describe the problem. The rescue captains are to locate lost ships using *question radar* directed to the lost ship captain, asking if point locations on the grid match the point locations of the lost ship corners. A sample of possible questions might include, "Is your lost ship located at (1, 1)?" "Is your lost ship located at (5, 2)?" The lost ship captain would reply *yes* or *no*.

5. While the rescue captains are using their *question radar*, direct them to record the location points on their blank grids. Have them place a small *x* on points that match the vertices of the lost ship and an *o* on points that are a *miss*. Demonstrate the procedure using the overhead projector.

6. When they think they have located enough points to establish the location and position of the lost ships, have them connect the *x* points to discover the shape and position of the rescued ships. The rescue team captains should then announce that they have located the _____ ships (giving the name of the geometric figures) at points,_____, _____, _____. (Example: triangle ship at (5, 2), (4, 4), and (2, 2).) The lost ship captain responds with either a *rescued* or *missed* response.

7. The game continues until the lost ship has been rescued. Repeat the procedures with partners switching roles. The students will need a new grid sheet each time they play the game.

8. Play the game several times. For variety, instead of allowing the students to illustrate the shapes, have them draw a polygon from the box you prepared. Have them place their shape with the corners at points on the grid.

Connecting Learning

1. Describe the problem you needed to solve.
2. How did you begin your search for the lost ship?
3. Describe anything you discovered that helped you find more *yes* responses than *no* responses from the lost ship captain.
4. Describe and name the shape of the ship you rescued. How many corners and sides did the shape have?
5. Describe the shape of the ship you rescued using vocabulary such as intersecting and parallel lines.
6. Describe and name the shape of the ship you drew. How many corners and sides did the shape have?
7. How many points did you need to use to draw a triangular ship?
8. How many points did you need to use to draw a rectangular ship?
9. How many points would you need to use to draw a ship that has the shape of an octagon?
10. When you were asking for locations on the grid, which axis was it necessary to name first? Why do you need to name the same axis first each time?
11. How did the coordinate grid help you identify the location of the shapes?
12. Describe how you might change your rescue plan the next time you search for a lost ship.

Extensions

1. Give the class a chance to explore many different geometric shapes such as pentagons, octagons, and other polygons. Have them name the points.
2. Challenge the students to build shape ships with four sides that are unequal in length, or two sides equal and two unequal.
3. Have the students draw other shapes on a coordinate grid, such as buildings, objects, etc. Have the students name the shapes used to illustrate these. Ask the students to identify the location of these objects by naming the points of all the corners of the drawn objects.

Curriculum Correlation

Glass, Julie. *The Fly on the Ceiling, A Math Myth*. Random House. New York. 1998.

A story about how the very messy French philosopher, René Descartes (portrayed as Ren Descartes in the book), invented an ingenious way to keep track of his possessions using a coordinate grid.

* Reprinted with permission from *Principles and Standards for School Mathematics,* 2000 by the National Council of Teachers of Mathematics. All rights reserved.

Ship Shape

Key Question

How can you tell what the shape of the hidden ship is?

Learning Goals

Students will:

- locate the position of shapes on a coordinate grid,
- apply their knowledge of attributes of various shapes by locating shapes on a coordinate grid, and
- use problem-solving strategies to locate shapes on a coordinate grid.

Ship Shape Grid

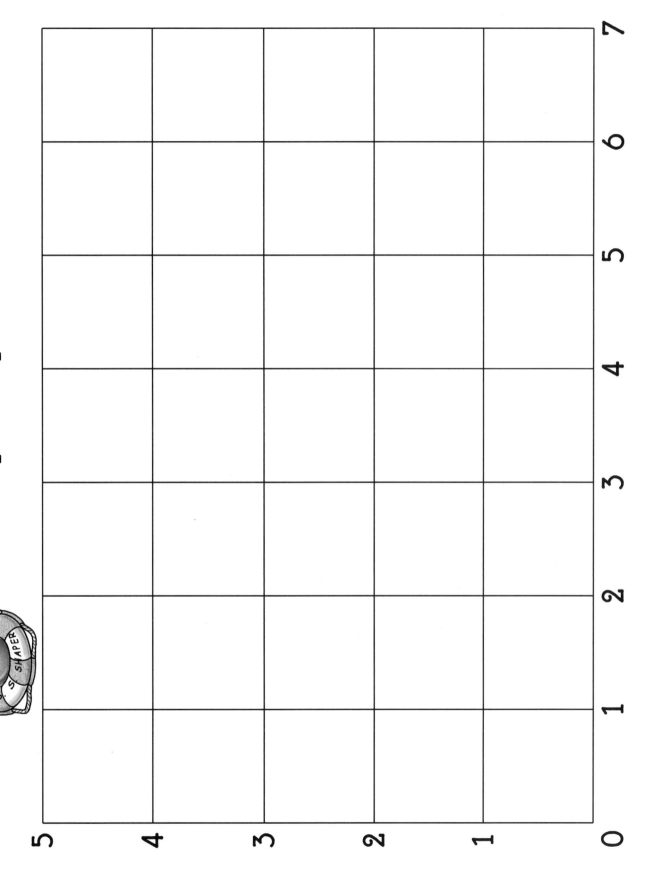

SHAPES, SOLIDS, AND MORE 371 © 2009 AIMS Education Foundation

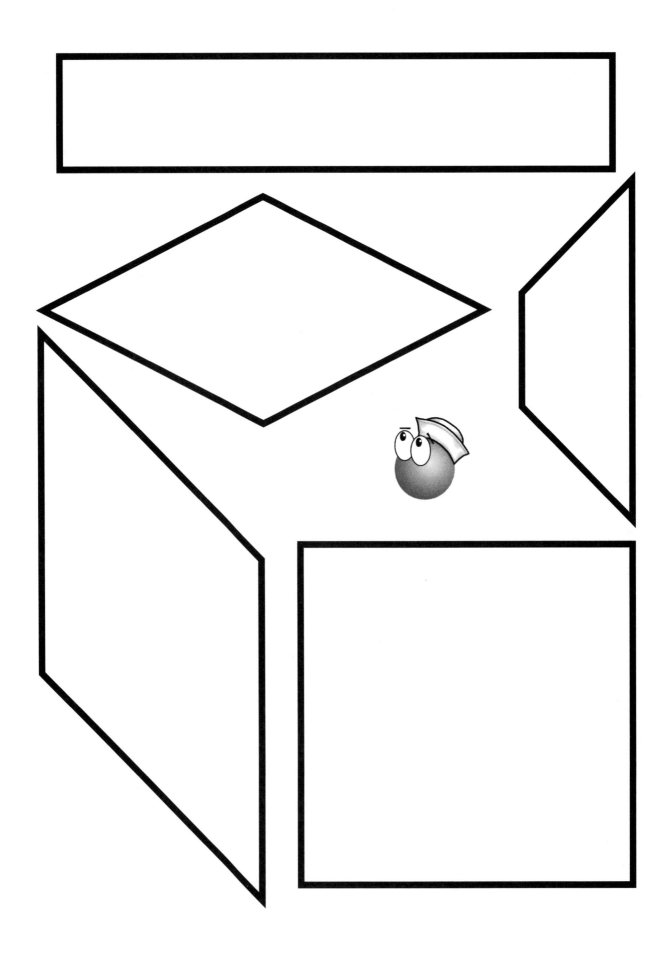

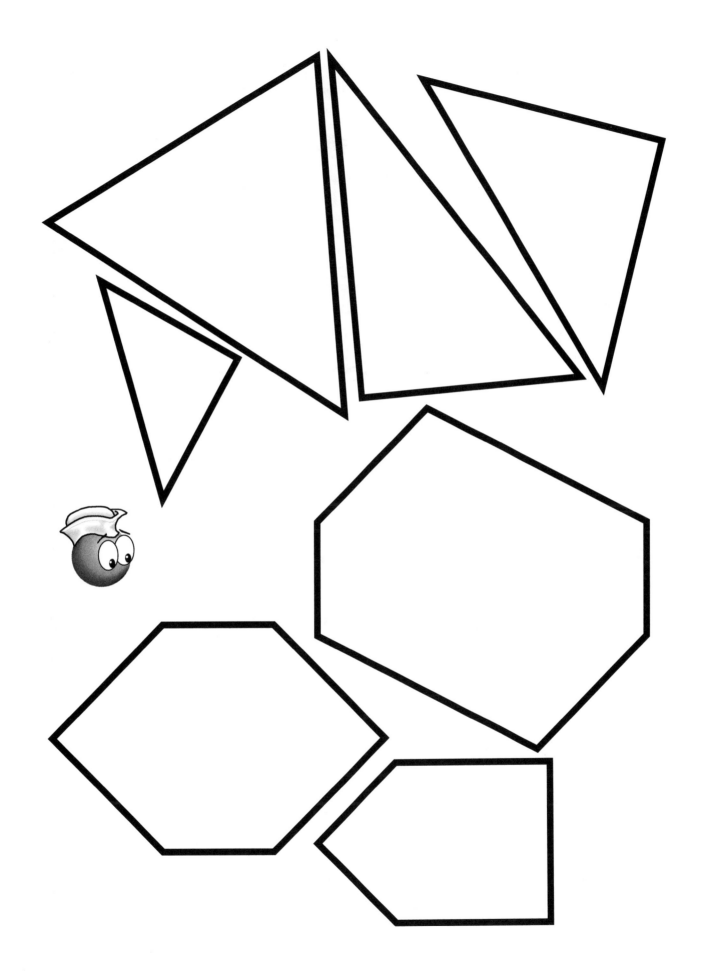

SHAPES, SOLIDS, AND MORE 373 © 2009 AIMS Education Foundation

Ship Shape

Connecting Learning

1. Describe the problem you needed to solve.

2. How did you begin your search for the lost ship?

3. Describe anything you discovered that helped you find more *yes* responses than *no* responses from the lost ship captain.

4. Describe and name the shape of the ship you rescued. How many corners and sides did the shape have?

5. Describe the shape of the ship you rescued using vocabulary such as intersecting and parallel lines.

6. Describe and name the shape of the ship you drew. How many corners and sides did the shape have?

Connecting Learning

7. How many points did you need to use to draw a triangular ship?

8. How many points did you need to use to draw a rectangular ship?

9. How many points would you need to use to draw a ship that has the shape of an octagon?

10. When you were asking for locations on the grid, which axis was it necessary to name first? Why do you need to name the same axis first each time?

11. How did the coordinate grid help you identify the location of the shapes?

12. Describe how you might change your rescue plan the next time you search for a lost ship.

Object of the Game
To have four boxes in a row, column, or diagonal covered with chips

Materials
For each student:
  one *Ge-O* game board
  covering chips *(see Management 1)*

For the class:
  set of *Ge-O Clue Cards (see Management 3)*

Management
1. The *Ge-O* game boards can be duplicated on cardstock and laminated for extended use.
2. Each student will need 16 chips to cover their game board. Beans, bingo chips, or even small squares of paper will work.
3. Duplicate the *Ge-O Clue Cards* on card stock and laminate them for extended use. They will then need to be cut out. There are 13 clues for 16 spaces. There are two triangles, two intersecting lines, (one of which is intersecting line segments), and two lines (one of which is a line segment). Students may choose either one of the two to cover when the appropriate clue is given. This enables them to use some problem-solving strategies to win.

Procedure
1. Give a *Ge-O* gameboard and 16 chips to each student. Inform students of the game's objective.
2. Explain that you have a set of clue cards that describes shapes and lines. Tell the students that you will choose a card and will read the description out loud. When the students can identify the shape or lines you are describing, have them cover the box that matches the description with a chip. Tell them that when they have four boxes covered in a row, column, or diagonal, they are to call out "Ge-O."
3. Have the student that called out "Ge-O" tell which four shapes or lines were covered. Check to make sure the answers are correct.

SHAPES, SOLIDS, AND MORE

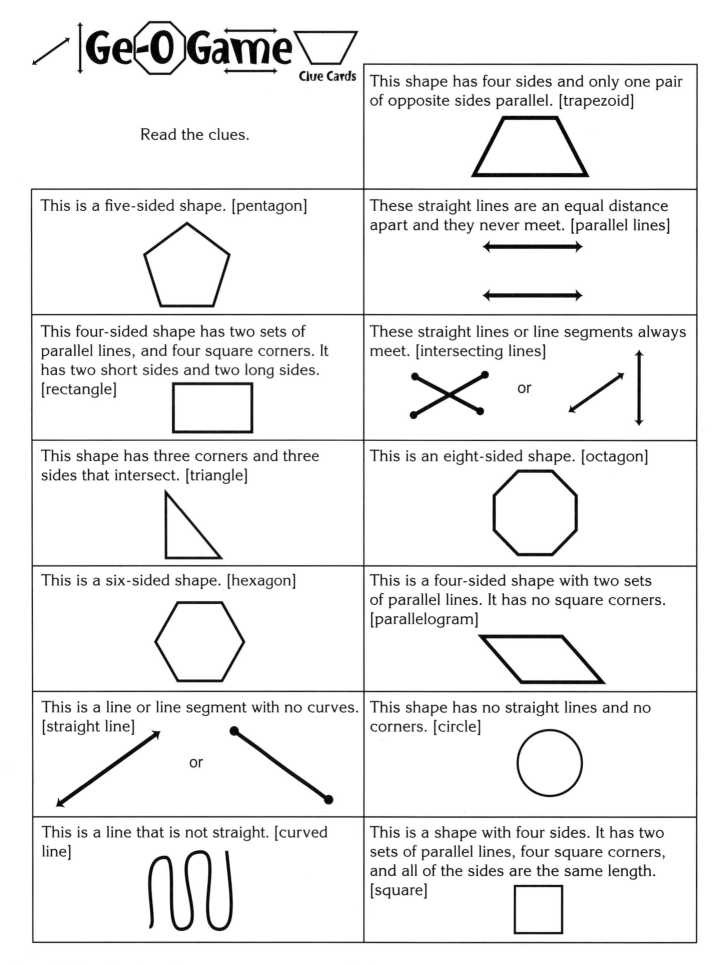

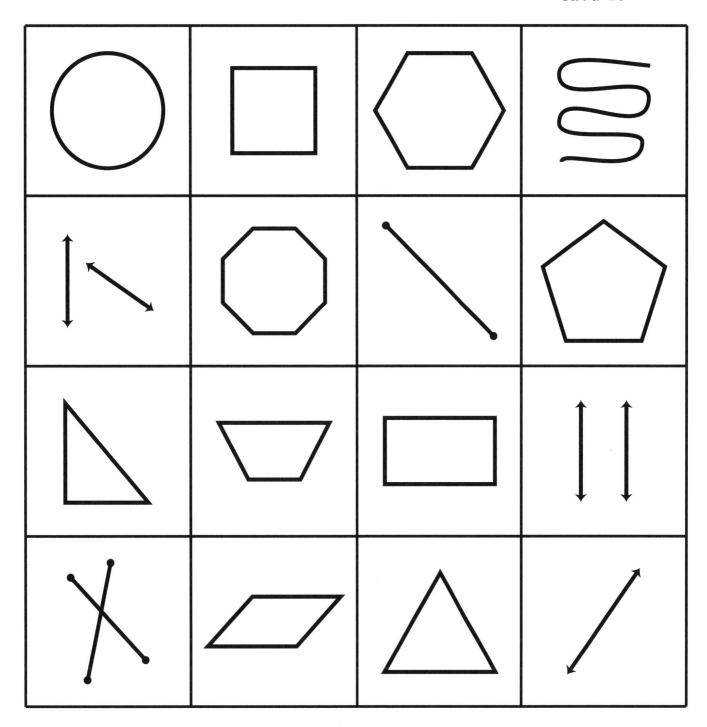

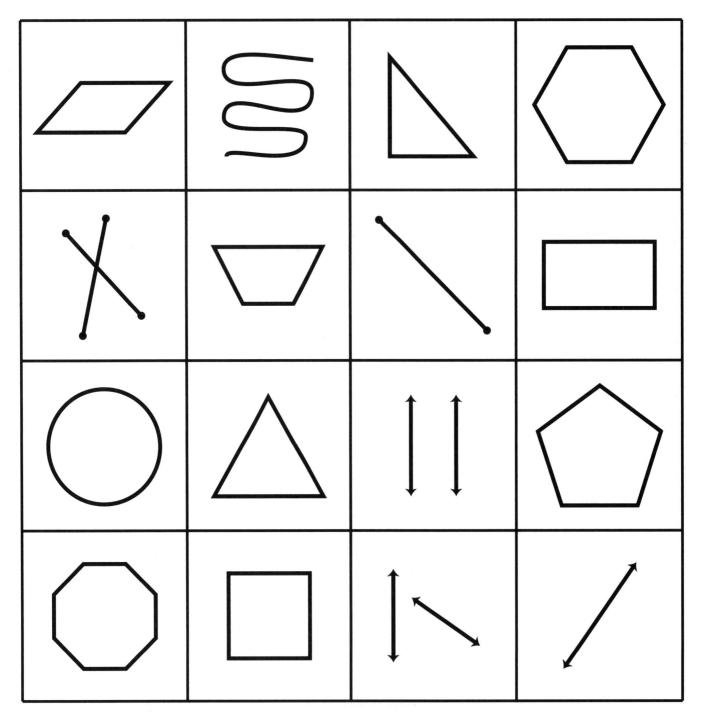

SHAPES, SOLIDS, AND MORE © 2009 AIMS Education Foundation

SHAPES, SOLIDS, AND MORE 381 © 2009 AIMS Education Foundation

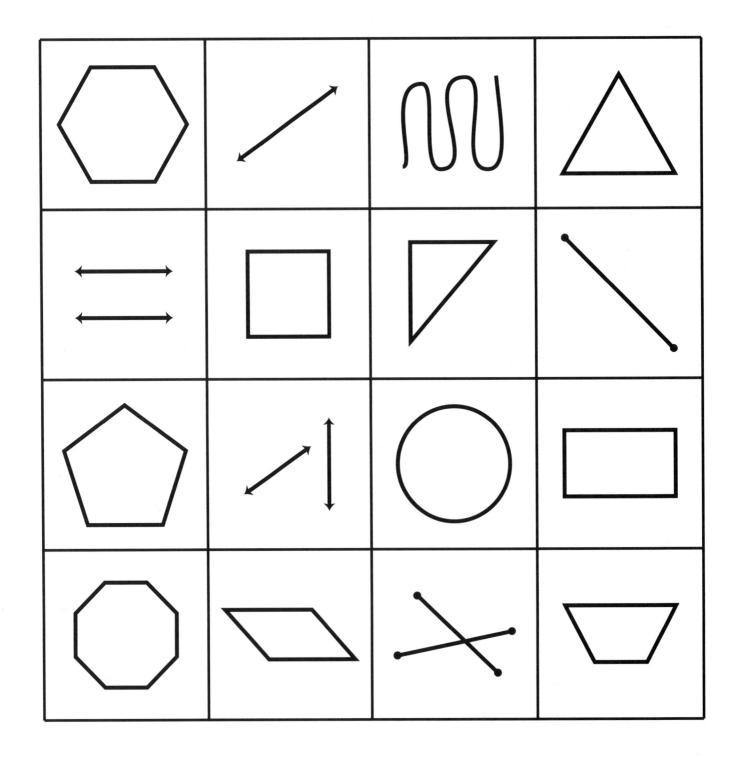

Card D

Who Has... Geometry

Students benefit from revisiting basic skills through repeated experiences in a variety of formats. While the adage *practice makes perfect* makes sense to us, there is considerable support for the idea that "drilling" for periods longer than 10 minutes a day may be counterproductive.

This learning game provides playful and intelligent practice within a very short period of time. The game features:
- an element of playfulness,
- minimum teacher preparation,
- time efficiency,
- mental stimulation and exercise, and
- student interest and motivation.

Who Has...? learning games use a statement/question format. A whole class experience may be facilitated by a teacher, a student, or a classroom assistant. Students may participate individually or in pairs.

Games can be played with the following cards:
- Use only cards 1-12.
- Use only cards 13-23.
- Use cards 1-11 and 24.

Management
1. Copy the cards onto card stock. Color the shapes as indicated.
2. Laminate the cards for extended use and cut out.

Procedure
1. Distribute one card to each student or pair of students.
2. Begin with any card other than numbers 12 and 23.
3. Ask a student to begin by reading his or her card aloud. (For example: "I have a red circle. Who has a green rectangle?")
4. Direct the student holding the answer to the card to respond by reading aloud the card. (For example: "I have a green rectangle. Who has a blue square?")
5. Continue the game until the cycle returns to the beginning card. (In this example, "I have a red circle.")

Who Has... Geometry Key

1. I have a red circle. Who has a green rectangle?
2. I have a green rectangle. Who has a blue square?
3. I have a blue square. Who has a yellow triangle?
4. I have a yellow triangle. Who has an orange rectangle?
5. I have an orange rectangle. Who has a green triangle?
6. I have a green triangle. Who has a blue circle?
7. I have a blue circle. Who has a purple triangle?
8. I have a purple triangle. Who has an orange square?
9. I have an orange square. Who has a yellow circle?
10. I have a yellow circle. Who has a blue triangle?
11. I have a blue triangle. Who has a brown rectangle?
12. I have a brown rectangle. Who has a red circle?
13. I have a brown rectangle. Who has a yellow parallelogram?
14. I have a yellow parallelogram. Who has a blue trapezoid?
15. I have a blue trapezoid. Who has a red rhombus?
16. I have a red rhombus. Who has a green pentagon?
17. I have a green pentagon. Who has an orange octagon?
18. I have an orange octagon. Who has a purple hexagon?
19. I have a purple hexagon. Who has parallel lines.
20. I have parallel lines. Who has a yellow pentagon?
21. I have a yellow pentagon. Who has intersecting lines?
22. I have intersecting lines. Who has an orange parallelogram?
23. I have an orange parallelogram. Who has a brown rectangle?
24. I have a brown rectangle. Who has a red circle?

SHAPES, SOLIDS, AND MORE

I have a

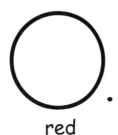
red

Who has a

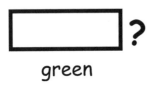
green

I have a

green

Who has a

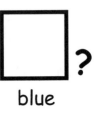
blue

I have a

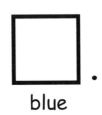
blue

Who has a

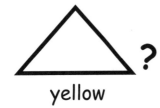
yellow

I have a

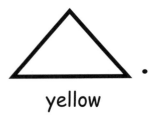
yellow

Who has an

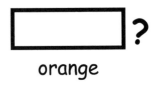
orange

SHAPES, SOLIDS, AND MORE

I have an .
orange

Who has a 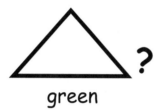?
green

I have a 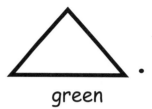.
green

Who has a 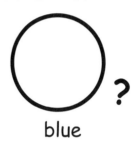?
blue

I have a 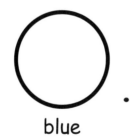.
blue

Who has a 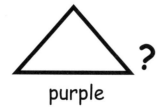?
purple

I have a 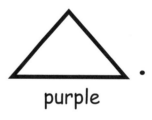.
purple

Who has an 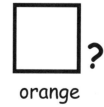?
orange

I have an 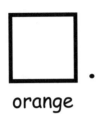.
orange

Who has a 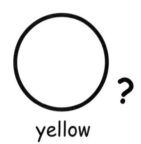?
yellow

I have a 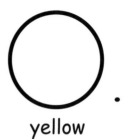.
yellow

Who has a 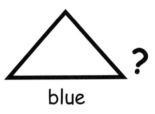?
blue

I have a 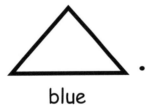.
blue

Who has a 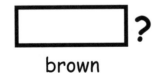?
brown

I have a 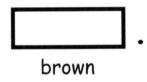.
brown

Who has a 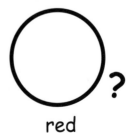?
red

I have a

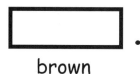
brown

Who has a

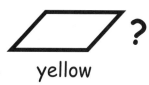
yellow

I have a

yellow

Who has a

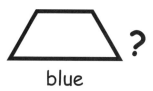
blue

I have a

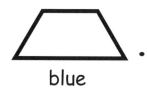
blue

Who has a

red

I have a

red

Who has a

green

I have a .
green

Who has an ?
orange

I have an .
orange

Who has a ?
purple

I have a .
purple

Who has 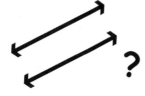?

I have 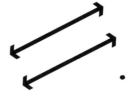.

Who has a 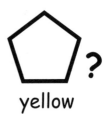?
yellow

I have a .
yellow

Who has ?

I have .

Who has an ?
orange

I have an 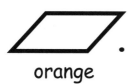.
orange

Who has a 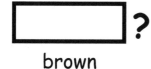?
brown

I have an 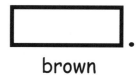.
brown

Who has a ?
red

SHAPES, SOLIDS, AND MORE

Triple Treasure Trivia

Topic
Properties of 2-D figures and 3-D objects

Key Question
What do the three words on each card have in common?

Learning Goal
Students will identify the relationship between three items.

Guiding Documents
Project 2061 Benchmarks
- Numbers and shapes—and operations on them—help to describe and predict things about the world around us.
- Shapes such as circles, squares, and triangles can be used to describe many things that can be seen.
- Circles, squares, triangles, and other shapes can be found in nature and in things people build.

NCTM Standard 2000*
- Identify, compare, and analyze attributes of two- and three-dimensional shapes and develop vocabulary to describe the attributes

Math
Geometry
    2-D shapes
    3-D solids

Integrated Processes
Observing
Relating

Materials
Game cards
Game pieces
Game board
Bell, optional

Background Information
    There is a commercial game on the market that challenges players to identify a common bond between three words, people, places, etc. For example, what do the words *element*, *odyssey*, and *pilot* all have in common? (All are models of cars currently manufactured by Honda.) This activity uses this same idea to develop students' geometric vocabularies.
    The game consists of 20 cards, each with three words or phrases. All of the words or phrases have a two- or three-dimensional shape in common. For example, a penny, a DVD, and a paper plate are all circular, while a stick of butter, a box of crayons, and a file cabinet are all rectangular prisms.

Management
1. Divide your class into four teams.
2. Copy the clue cards onto card stock and cut them out.
3. Make a transparency of the game board to place on the overhead.
4. Each team will need a small object to use as a playing piece. If desired, collect four bells or other items that students can use to "ring in" to answer.
5. Depending on the ages and abilities of your students, you may wish to remove some of the cards.

Procedure
1. Tell the students that they will be playing a thinking game. Explain that they will be divided into teams and that each team will be shown a card with three items and that it will be their job to determine what the three items have in common.
2. The rules for *Triple Treasure Trivia* follow this teacher's text.
3. Explain the object and rules of the game and allow the students to play.
4. When a winning team has been determined or time is up, discuss what types or common bonds the items shared. Invite students to make suggestions for additional card sets.

Connecting Learning
1. Was it harder to figure out two-dimensional or three-dimensional objects? Why do you think that is?
2. Which clue was the easiest? Why?
3. What other objects could we list on a cylinder card? ... triangle card? ... etc.?
4. If you could recreate this game, what would it be like?

Solutions
Responses in the two-dimensional shapes category include: circles, squares, triangles, rectangles, and pentagons.

Responses in the three-dimensional solids category include: cubes, spheres, cones, cylinders, and rectangular solids.

* Reprinted with permission from *Principles and Standards for School Mathematics,* 2000 by the National Council of Teachers of Mathematics. All rights reserved.

Triple Treasure Trivia Rules

Object: To determine what the three words on each card have in common

Players: Four teams, with one "active player" representing each team for each question. This active player must change for every question until each person has had a chance to respond. At that point, the rotation of active players begins again in the same order.

Play:

1. Shuffle the cards and place them face down in a pile. Select the top card from the pile.

2. Ask the active player from each team to identify him or herself.

3. Read the three words on the card. The first active player to "ring in" gets to respond. You must have finished reading all three words before players can ring in.*

4. If the active player is correct, that team advances one space toward the finish. If the first active player is incorrect, the next active player to "ring in" gets to respond. If this player is also incorrect, the next player gets to respond, and so on. If none of the active players are correct, the card is forfeited and no one gets to move. New active players respond to the next question.

5. Play continues in this fashion until one team reaches the finish.

* You can use a variety of methods to have students "ring in." Using small bells (the sort found in libraries) is perhaps the easiest method. You can also provide empty metal coffee cans and wooden spoons, or simply have students raise their hands.

Triple Treasure Trivia

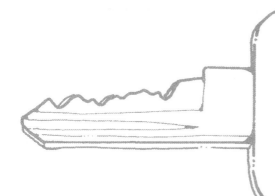

Key Question

What do the three words on each card have in common?

Learning Goal

identify the relationship between three items.

Answer: All are circles	**Clues:** Penny / DVD / Paper plate	**Answer:** All are rectangles	**Clues:** License plate / Birthday card / Postage stamp
Answer: All are circles	**Clues:** Tortilla / Ritz® Cracker / Oreo® Cookie	**Answer:** All are rectangles	**Clues:** American flag / Picture frame / Window
Answer: All are circles	**Clues:** Clock / Wedding ring / Hula hoop	**Answer:** All are rectangles	**Clues:** Piece of paper / Basketball court / Chalkboard/Whiteboard
Answer: All are triangles	**Clues:** Doritos® tortilla chip / Sailboat sail / Slice of pizza	**Answer:** All are squares	**Clues:** Sheet of origami paper / Sandwich baggie / Saltine cracker
Answer: All are triangles	**Clues:** Face of a pyramid / Arrowhead / Half a sandwich	**Answer:** All are squares	**Clues:** Checkerboard space / Post-it™ note / American cheese slice

Answer: All are pentagons	**Clues:** Home plate The black shapes on a soccer ball The outline of a house	Answer: All are cones	**Clues:** Party hat Sno-cone cup Teepee
Answer: All are cubes	**Clues:** Dice Rubik's Cube Alphabet block	Answer: All are spheres	**Clues:** Basketball Globe Bubbles
Answer: All are cylinders	**Clues:** Soda can Toilet paper tube Piece of chalk	Answer: All are spheres	**Clues:** Soccer ball Marble Orange
Answer: All are cylinders	**Clues:** Roll of wrapping paper Straw Candle	Answer: All are rectangular solids/prisms	**Clues:** Stick of butter Box of crayons File cabinet
Answer: All are cylinders	**Clues:** Glue stick AA Battery Tube of lipstick	Answer: All are rectangular solids/prisms	**Clues:** Shoe box Whiteboard eraser Brick

SHAPES, SOLIDS, AND MORE

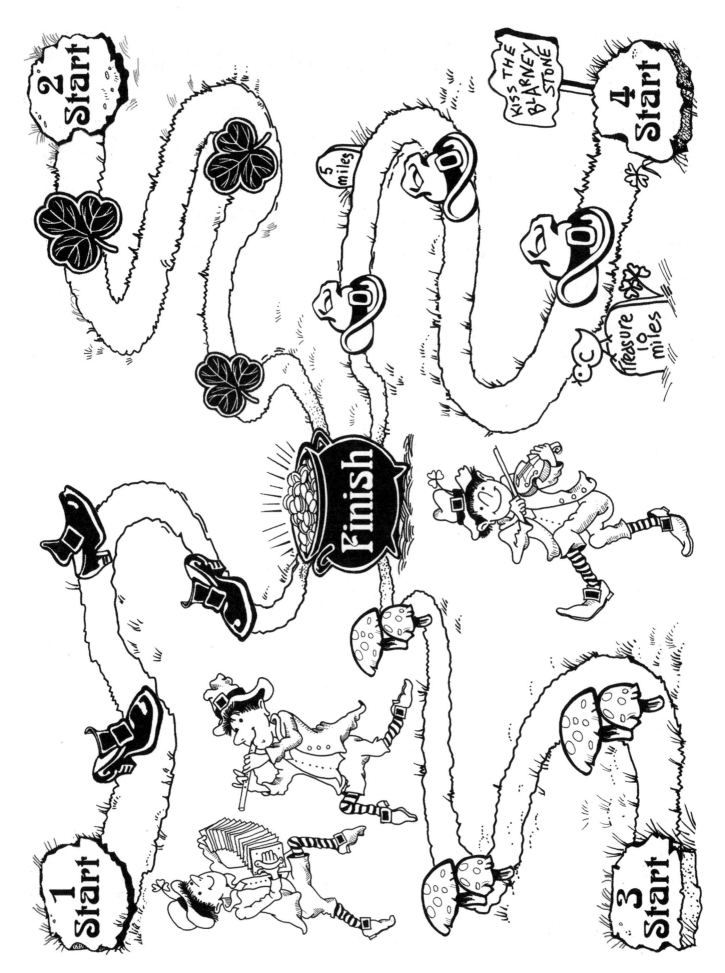

Triple Treasure Trivia

Connecting Learning

1. Was it harder to figure out two-dimensional or three-dimensional objects? Why do you think that is?

2. Which clue was the easiest? Why?

3. What other objects could we list on a cylinder card? … triangle card?

4. If you could recreate this game what would it be like?

Word Search

Topic
Geometric vocabulary

Key Question
What words help us describe two-dimensional shapes?

Learning Goal
Students will use geometric vocabulary to describe two-dimensional shapes.

Guiding Document
NCTM Standards 2000
* *Describe attributes and parts of two- and three-dimensional shapes*
* *Recognize, name, build, draw, compare, and sort two- and three-dimensional shapes*

Math
Geometry
    2-D shapes
        properties
        vocabulary

Integrated Processes
Observing
Classifying
Comparing and contrasting
Relating

Materials
One brown paper shopping bag
Pocket chart (see *Management 1*)
One set of geometry vocabulary cards
    (see *Management 4*)
Several two-dimensional shapes (see *Management 5*)
Lined writing paper

Background Information
This activity can be used to reinforce previously taught lessons on the naming and describing of two-dimensional shapes or as an assessment of your students' use of proper vocabulary.

Management
1. If a pocket chart is not available, the marker tray will work.
2. Prior experience with two-dimensional shapes is suggested before beginning this lesson.
3. This activity should be introduced as a whole group activity. It could later be moved to a center for additional practice.
4. Prior to the lesson, prepare the geometry vocabulary cards. A suggested list of terms that can be written on sentence strips and used in a pocket chart has been included. Add any other shapes and attributes that you and your students have explored.
5. Prepare your shopping bag by placing several shapes that are 10 centimeters or larger inside. These shapes can be cut out of card stock, vinyl placemats, or craft foam using a die cut machine.

Procedure
Part One
1. Display the geometry vocabulary cards by taping them to the board.
2. Tell the students that they are going to select sentence strips that name and describe geometric shapes. Inform them that the strips will be placed in the pocket chart.
3. Have one student remove a shape from the paper bag and put it into the pocket chart. Urge the student to find as many words as he or she can to describe and name the shape. Invite the student to put the words in the pocket chart under the shape.

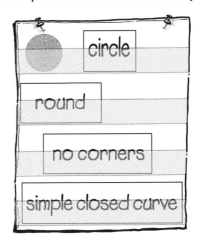

4. Discuss the vocabulary that was chosen. Be sure that the vocabulary correctly matches the shape and that no attributes were left out.
5. Repeat the procedure until several children have had an opportunity to match the attributes with the shapes.

Part Two
1. Review with students the differences between the name of the two-dimensional shapes and the attributes of the two-dimensional shapes. (Triangle is a name, its attributes are three sides, three corners, etc.)

SHAPES, SOLIDS, AND MORE   © 2009 AIMS Education Foundation

2. After the students have matched their shapes with the name and attributes, ask them to write a story about their shape and include the attributes.

Connecting Learning
1. How are triangles alike? …different?
2. What shape has five sides?
3. What shapes have four corners?
4. Name a shape that has parallel sides.
5. Name a shape that has intersecting lines.
6. When we describe a shape, what do we look at? [the number of sides, number of corners, and lines that make up the sides]
7. Were some shapes easier to describe than others? Explain.
8. How are squares and rectangles alike? …different?

Suggested Geometry Vocabulary
square
circle
triangle
rectangle
parallelogram
rhombus
trapezoid
pentagon
hexagon
octagon
polygon
lines
rays
parallel lines
intersecting lines
perpendicular lines
simple closed curve
round
no corners
3 corners
4 corners
5 corners
6 corners
8 corners
square corners
3 sides
4 sides
5 sides
6 sides
8 sides

---

* Reprinted with permission from *Principles and Standards for School Mathematics,* 2000 by the National Council of Teachers of Mathematics. All rights reserved.

Word Search

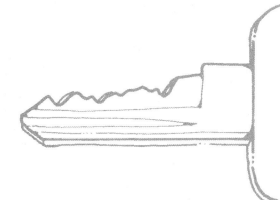

Key Question

What words help us describe two-dimensional shapes?

Learning Goal

use geometric vocabulary to describe two-dimensional shapes.

Word Search

Connecting Learning

1. How are triangles alike? ...different?

2. What shape has five sides?

3. What shapes have four corners?

4. Name a shape that has parallel sides.

5. Name a shape that has intersecting lines.

6. When we describe a shape, what do we look at?

Word Search

Connecting Learning

7. Were some shapes easier to describe than others? Explain.

8. How are squares and rectangles alike? ...different?

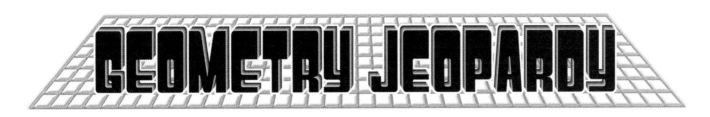

Topic
Geometric vocabulary

Key Question
How can playing a game help us review geometry terms?

Learning Goal
Students will review geometry terms through a game of jeopardy.

Guiding Document
NCTM Standards 2000*
- *Recognize, name, build, draw, compare, and sort two- and three-dimensional shapes*
- *Describe attributes and parts of two- and three-dimensional shapes*
- *Identify, compare, and analyze attributes of two- and three-dimensional shapes and develop vocabulary to describe the attributes*
- *Classify two- and three-dimensional shapes according to their properties and develop definitions of classes of shapes such as triangles and pyramids*

Math
Geometry
    vocabulary

Integrated Processes
Observing
Comparing and contrasting
Drawing conclusions
Applying

Materials
Die
3" x 5" cards
Clock with a second hand
Ring in devices (see *Management 4*)

Background information
In order for students to become comfortable with geometric terms, it is important for them to frequently use and review them. This activity will provide a playful game format for students to use and apply this vocabulary. It will also allow you as the teacher to assess their understanding of the terms. The review game the students will be playing is *Geometry Jeopardy*. Unlike regular Jeopardy, students will not have to answer in the form of a question. They will, however, earn or lose points based on correct answers, so their points will always be in jeopardy.

Management
1. Prior to playing *Geometry Jeopardy*, make a six by six grid on the wall or board. Label the categories *Triangles*, *Circles*, *Polygons*, *Lines and Angles*, *Symmetry and Transformations*, and *Solids*.
2. To prepare point/question cards, copy, cut, and glue the questions included in this activity onto one side of the 3" x 5" cards. On the opposite side, use a marker to write the point value that matches each question.
3. Divide the class into two teams or multiple small teams.
4. Gather one "ring in" device for each team. Whistles, hotel bells, chimes, etc., all work well.
5. Prior to playing the game, arrange the classroom so that there is a contestant's table with enough spots for one representative from each team to stand. This table should be positioned so that contestants have a clear view of the game board and have space behind for the rest of the team members.
6. You may choose to end the game with a "Final Jeopardy" question. This will allow teams to decide how many points they would like to wager on one final question. A possible Final Jeopardy question might be, "What is the movement of a figure by slides (translations), flips (reflections), or turns (rotations) called?" [transformation]

Procedure
1. Tell the class that they will be spending today's math class reviewing geometric terms by playing a game of *Geometry Jeopardy*. Explain the following rules to students:
   - Each team will roll a die. The team with the highest number goes first.
   - Teams will stand in single file lines behind the contestant's table. The first player in each line will move forward to the table. The first player selects one of the five categories displayed on the upper row of the grid, and a point value. For example, if the player chooses circles for 300, the question would be about circles and the amount of points that could be lost or won would be 300. (Inform students that generally

the higher the point value, the more difficult the question.) The teacher removes the card and reads the question on the back. The first student to ring in gets 30 seconds to answer the question. (Adjust the time to the knowledge level of your students.) If two or more players ring in at the same time, players roll the die; the player with the highest number answers the question.
- A correct answer will add points to the team's score, and an incorrect answer subtracts point from the team. The card is then removed and if the player answered correctly, the next contestant in his/her line may choose the next category and point value. Teammates will rotate after each question. Play continues until the board is cleared or the allotted time is up. The team with the most points wins.

Connecting Learning
1. Which questions were the hardest to answer? Why?
2. Why is it important to know geometry terms?
3. What other ways could we review our geometry vocabulary?
4. What strategies did you use when deciding what category and point value to choose?

* Reprinted with permission from *Principles and Standards for School Mathematics*, 2000 by the National Council of Teachers of Mathematics. All rights reserved.

Jeopardy Questions

Solids

100
What is the name for a solid figure that is shaped like a can? [cylinder]

200
What is the name for a solid figure with a triangular base and triangular sides that meet at a point? [triangular pyramid]

300
What is the name for a solid figure with a square base and triangular sides that meet at a point? [square pyramid]

400
What is the name for a solid figure with six congruent square faces? [cube]

500
What is the name for a solid figure in which all six faces are rectangles? [rectangular prism, cube]

Polygons

100
What is the name for a polygon with three sides? [triangle]

200
What is the name for a six-sided shape? [hexagon]

300
How many sides does a quadrilateral have? [four]

400
What geometric figures are represented by the following signs: stop sign, speed limit sign, yield sign? [octagon, rectangle, triangle]

500
What are the names of the quadrilaterals with four equal sides? [square, rhombus]

Circles

100
If the radius of a circle is two, how long is the diameter? [Four. The radius is half the length of the diameter.]

200
What is the radius of a circle? [The radius is a line segment with an end point at the center of the circle and a point on the circle. It is half the diameter of the circle.]

300
How do you determine the circumference of a circle? [The circumference is the measure around the outside of the circle.]

400
Are circles polygons? Why or why not? [Circles are not polygons. They are curved; they are not made up of straight lines.]

500
What is the relationship between the radius and the diameter of a circle? [The radius is half the length of the diameter, or the diameter is twice the length of the radius.]

Lines and Angles

100
What are parallel lines? [lines that are equal distance apart at all points]

200
What is the name for lines that cross each other? [intersecting lines]

300
What is the name for lines that intersect to form four right angles? [perpendicular lines]

400
To make an obtuse angle from a right angle, which way would one of the rays have to move? [outward to make an obtuse angle because an obtuse angle is greater than 90 degrees]

500
Which of the following have perpendicular lines—right angles, obtuse angles, or acute angles? [right]

Symmetry and Transformations

100
What is another name for a reflection? [a flip]

200
Do all figures have a line or lines of symmetry? [No.]

300
What is another name for a translation? [a slide]

400
What are transformations? [when an object changes position and/or orientation either by flipping, turning, or sliding]

500
How many lines of symmetry does a square have? [4]

Triangles

100
How many sides does a triangle have? [3]

200
What is the name for a triangle with a 90-degree angle? [a right triangle]

300
What is the name for a triangle with three congruent sides? [an equilateral triangle]

400
What is the name for a triangle with no congruent sides? [a scalene triangle]

500
What is the name for a triangle with two congruent sides? [an isosceles triangle]

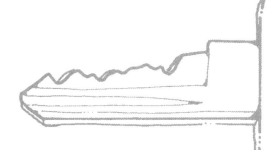

Key Question

How can playing a game help us review geometry terms?

Learning Goal

review geometry terms through a game of jeopardy.

Connecting Learning

1. Which questions were the hardest to answer? Why?

2. Why is it important to know geometry terms?

3. What other ways could we review our geometry vocabulary?

4. What strategies did you use when deciding what category and point value to choose?

Shape Draw

Purpose of the Game
Students will play a version of tic-tac-toe by drawing a 3-D shape out of a sock and matching it to either its corresponding name or picture.

Materials
For each pair of students:
  2 socks
  attribute beads in two colors (cylinders, spheres, and cubes)
  game board
  2 plastic cups, 9 oz

Management
1. This game is designed to be played in pairs. There are two versions of the game. Each version has the same rules, but uses a different game board. In *Version One,* the students will be matching 3-D attribute beads—cylinders, spheres, and cubes—to pictures of the shapes. In *Version Two,* the students will be matching the 3-D models to the name of each shape.
2. Both game boards have been placed on the same sheet of paper, but to avoid confusion, they should be cut apart before being given to students. Each pair of students will need one copy of the game board for each version.
3. Place a wide-mouth plastic cup in the toe of each sock. Fill each cup with the attribute beads that correspond to the spaces on the boards. Each student should have a sock filled with attribute beads in a different color than their partners. This will make it easy to tell what plays were made by each student.

Rules
1. One player begins by reaching into the sock and pulling out an attribute bead. The player must place the shape on the corresponding picture (or name) of the shape. Any empty square with that shape (name) may be covered by the student. If there are no empty squares with that shape (name), the student forfeits that turn, and the other player gets to draw.
2. Players take turns choosing shapes and covering squares until one player has three squares in a row horizontally, vertically, or diagonally.
3. If neither player is able to get three squares in a row, the game is a draw.

SHAPES, SOLIDS, AND MORE © 2009 AIMS Education Foundation

Shape Draw
Version One

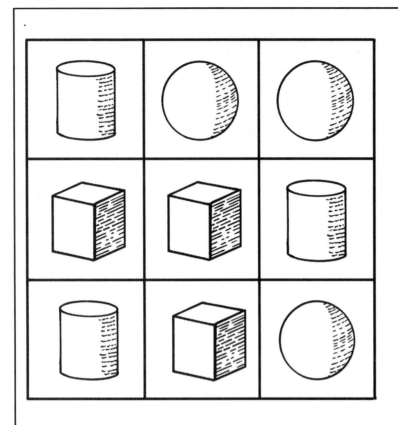

Shape Draw
Version Two

Cube	Sphere	Cylinder
Cylinder	Sphere	Sphere
Cube	Cylinder	Cube

Literature for Geometry Grades 2-3

Adler, David. *Shape Up! Fun with Triangles & Other Polygons.* Holiday House. New York. 1998.

Axelrod, Amy. *Pigs on the Ball: Fun With Math and Sports.* Simon & Schuster Children's Publishing. New York. 1998.

Bang, Molly. *The Paper Crane (Reading Rainbow Book).* HarperTrophy. New York. 1987.

Burns, Marilyn. *The Greedy Triangle.* Scholastic, Inc. New York. 1994.

Dodds, Dayle Ann. *The Shape of Things.* Scholastic, Inc. New York. 1999.

Emberley, Ed. *Ed Emberley's Picture Pie: A Book of Circle Art.* L.B Kids. New York. 1984.

Ernst, Lisa Campbell. *Sam Johnson & the Blue Ribbon Quilt.* HarperTrophy. New York. 1992.

Friedman, Aileen. *The Cloak for a Dreamer.* Scholastic Press. New York. 1995.

Froman, Robert. *Angles are Easy as Pie.* T.Y. Crowell Co. New York. 1976.

Glass, Ruth. *The Fly on the Ceiling.* Random House. New York. 1998.

Greene, Rhonda Gowler, *When a Line Bends...A Shape Begins.* Houghton Mifflin. Boston. 2001.

Hoban, Tana *Cubes, Cones, Cylinders, & Spheres.* Greenwillow Books. New York. 2000.

Hoban, Tana. *Shapes, Shapes, Shapes.* Harper Trophy. New York. 1996.

Hoban, Tana. *So Many Circles, So Many Squares.* Greenwillow Books. New York. 1998.

Hoban, Tana. *Spirals, Curves, Fanshapes and Lines.* Greenwillow Books. New York. 1992.

Hutchins, Pat. *Changes, Changes.* Scholastic, Inc. New York. 1987.

Maccarone, Grace. *Three Pigs, One Wolf, and Seven Magic Shapes.* Scholastic, Inc. New York. 1997.

Martin, Jacqueline B. *Snowflake Bentley.* Houghton Mifflin. Boston. 1998.

Morgan, Sally. *Triangles and Pyramids.* Thomson Learning. New York. 1995.

Murphy, Stuart J. *Captain Invincible and the Space Shapes.* HarperCollins. New York. 2001.

Murphy, Stuart J. *Circus Shapes: Recognizing Shapes.* HarperCollins. New York. 1998.

Neuschwander, Cindy. *Mummy Math: An Adventure in Geometry.* Henry Holt & Co. New York. 2005.

Neuschwander, Cindy. *Sir Cumference & the First Round Table.* Charlesbridge Publishing. Watertown, MA. 1997.

Neuschwander, Cindy, *Sir Cumference and the Sword in the Cone: A Math Adventure.* Charlesbridge Publishing. Watertown, MA. 2003.

McMillan, Bruce. *Mouse Views: What the Class Pet Saw.* Holiday House. New York. 1994.

Onyefulu, Ifeoma. *Triangle for Adaora: An African Book of Shapes.* Frances Lincoln. Kentish Town, London. 2007.

Paul, A.W. *Eight Hands Round: A Patchwork Quilt.* HarperTrophy. New York. 1996.

Pilegard, Virginia Walton. *The Warlord's Puzzle.* Pelican Publishing Co., Inc. Gretna, LA. 2000.

Pluckrose, Henry. *Shape.* Children's Press. Chicago. 1995.

Rogers, Paul. *The Shapes Game.* Henry Holt & Co. New York. 1989.

Serfozo, Mary. *There's a Square: A Book About Shapes.* Scholastic, Inc. New York. 1996.

Seuss, Dr. *The Shape of Me and Other Stuff.* Random House. New York. 1973.

Tompert, Ann. *Grandfather Tang's Story.* Crown Publishers, Inc. New York. 1990.

The AIMS Program

AIMS is the acronym for "**A**ctivities **I**ntegrating **M**athematics and **S**cience." Such integration enriches learning and makes it meaningful and holistic. AIMS began as a project of Fresno Pacific University to integrate the study of mathematics and science in grades K-9, but has since expanded to include language arts, social studies, and other disciplines.

AIMS is a continuing program of the non-profit AIMS Education Foundation. It had its inception in a National Science Foundation funded program whose purpose was to explore the effectiveness of integrating mathematics and science. The project directors, in cooperation with 80 elementary classroom teachers, devoted two years to a thorough field-testing of the results and implications of integration.

The approach met with such positive results that the decision was made to launch a program to create instructional materials incorporating this concept. Despite the fact that thoughtful educators have long recommended an integrative approach, very little appropriate material was available in 1981 when the project began. A series of writing projects ensued, and today the AIMS Education Foundation is committed to continuing the creation of new integrated activities on a permanent basis.

The AIMS program is funded through the sale of books, products, and professional-development workshops, and through proceeds from the Foundation's endowment. All net income from programs and products flows into a trust fund administered by the AIMS Education Foundation. Use of these funds is restricted to support of research, development, and publication of new materials. Writers donate all their rights to the Foundation to support its ongoing program. No royalties are paid to the writers.

The rationale for integration lies in the fact that science, mathematics, language arts, social studies, etc., are integrally interwoven in the real world, from which it follows that they should be similarly treated in the classroom where students are being prepared to live in that world. Teachers who use the AIMS program give enthusiastic endorsement to the effectiveness of this approach.

Science encompasses the art of questioning, investigating, hypothesizing, discovering, and communicating. Mathematics is a language that provides clarity, objectivity, and understanding. The language arts provide us with powerful tools of communication. Many of the major contemporary societal issues stem from advancements in science and must be studied in the context of the social sciences. Therefore, it is timely that all of us take seriously a more holistic method of educating our students. This goal motivates all who are associated with the AIMS Program. We invite you to join us in this effort.

Meaningful integration of knowledge is a major recommendation coming from the nation's professional science and mathematics associations. The American Association for the Advancement of Science in *Science for All Americans* strongly recommends the integration of mathematics, science, and technology. The National Council of Teachers of Mathematics places strong emphasis on applications of mathematics found in science investigations. AIMS is fully aligned with these recommendations.

Extensive field testing of AIMS investigations confirms these beneficial results:
1. Mathematics becomes more meaningful, hence more useful, when it is applied to situations that interest students.
2. The extent to which science is studied and understood is increased when mathematics and science are integrated.
3. There is improved quality of learning and retention, supporting the thesis that learning which is meaningful and relevant is more effective.
4. Motivation and involvement are increased dramatically as students investigate real-world situations and participate actively in the process.

We invite you to become part of this classroom teacher movement by using an integrated approach to learning and sharing any suggestions you may have. The AIMS Program welcomes you!

AIMS Education Foundation Programs

When you host an AIMS workshop for elementary and middle school educators, you will know your teachers are receiving effective, usable training they can apply in their classrooms immediately.

AIMS Workshops are Designed for Teachers
- Correlated to your state standards;
- Address key topic areas, including math content, science content, and process skills;
- Provide practice of activity-based teaching;
- Address classroom management issues and higher-order thinking skills;
- Give you AIMS resources; and
- Offer optional college (graduate-level) credits for many courses.

AIMS Workshops Fit District/Administrative Needs
- Flexible scheduling and grade-span options;
- Customized (one-, two-, or three-day) workshops meet specific schedule, topic, state standards, and grade-span needs;
- Prepackaged four-day workshops for in-depth math and science training available (includes all materials and expenses);
- Sustained staff development is available for which workshops can be scheduled throughout the school year;
- Eligible for funding under the Title I and Title II sections of No Child Left Behind; and
- Affordable professional development—consecutive-day workshops offer considerable savings.

University Credit—Correspondence Courses
AIMS offers correspondence courses through a partnership with Fresno Pacific University.
- Convenient distance-learning courses—you study at your own pace and schedule. No computer or Internet access required!

Introducing AIMS State-Specific Science Curriculum
Developed to meet 100% of your state's standards, AIMS' State-Specific Science Curriculum gives students the opportunity to build content knowledge, thinking skills, and fundamental science processes.
- Each grade-specific module has been developed to extend the AIMS approach to full-year science programs. Modules can be used as a complete curriculum or as a supplement to existing materials.
- Each standards-based module includes mathreading, hands-on investigations, and assessments.

Like all AIMS resources, these modules are able to serve students at all stages of readiness, making these a great value across the grades served in your school.

For current information regarding the programs described above, please complete the following form and mail it to: P.O. Box 8120, Fresno, CA 93747.

Information Request

Please send current information on the items checked:

___ *Basic Information Packet* on AIMS materials
___ Hosting information for AIMS workshops
___ AIMS State-Specific Science Curriculum

Name: _____

Phone: _____ E-mail: _____

Address: _____
          Street   City   State   Zip

SHAPES, SOLIDS, AND MORE   © 2009 AIMS Education Foundation

Magazine

YOUR K-9 MATH AND SCIENCE CLASSROOM ACTIVITIES RESOURCE

The AIMS Magazine is your source for standards-based, hands-on math and science investigations. Each issue is filled with teacher-friendly, ready-to-use activities that engage students in meaningful learning.

- *Four issues each year (fall, winter, spring, and summer).*

Current issue is shipped with all past issues within that volume.

1823	Volume XXIII	2008-2009	$19.95
1824	Volume XXIV	2009-2010	$19.95
1825	Volume XXV	2010-2011	$19.95

Two-Volume Combination

M20810	Volumes XXIII & XXIV	2008-2010	$34.95
M20911	Volumes XXIV & XXV	2009-2011	$34.95

Complete volumes available for purchase:

1802	Volume II	1987-1988	$19.95
1804	Volume IV	1989-1990	$19.95
1805	Volume V	1990-1991	$19.95
1807	Volume VII	1992-1993	$19.95
1808	Volume VIII	1993-1994	$19.95
1809	Volume IX	1994-1995	$19.95
1810	Volume X	1995-1996	$19.95
1811	Volume XI	1996-1997	$19.95
1812	Volume XII	1997-1998	$19.95
1813	Volume XIII	1998-1999	$19.95
1814	Volume XIV	1999-2000	$19.95
1815	Volume XV	2000-2001	$19.95
1816	Volume XVI	2001-2002	$19.95
1817	Volume XVII	2002-2003	$19.95
1818	Volume XVIII	2003-2004	$19.95
1819	Volume XIX	2004-2005	$19.95
1820	Volume XX	2005-2006	$19.95
1821	Volume XXI	2006-2007	$19.95
1822	Volume XXII	2007-2008	$19.95

Volumes II to XIX include 10 issues.

Call 1.888.733.2467 or go to www.aimsedu.org

Subscribe to the AIMS Magazine

$19.95 a year!

AIMS Magazine is published four times a year.

Subscriptions ordered at any time will receive all the issues for that year.

AIMS Online—www.aimsedu.org

To see all that AIMS has to offer, check us out on the Internet at www.aimsedu.org. At our website you can search our activities database; preview and purchase individual AIMS activities; learn about state-specific science, college courses, and workshops; buy manipulatives and other classroom resources; and download free resources including articles, puzzles, and sample AIMS activities.

AIMS News

While visiting the AIMS website, sign up for AIMS News, our FREE e-mail newsletter.
Included in each month's issue you will find:

- Information on what's new at AIMS (publications, materials, state-specific science modules, etc.)
- A special money-saving offer for a book and/or product; and
- Free sample activities.

Sign up today!

AIMS Program Publications

Actions with Fractions, 4-9
The Amazing Circle, 4-9
Awesome Addition and Super Subtraction, 2-3
Bats Incredible! 2-4
Brick Layers II, 4-9
The Budding Botanist, 3-6
Chemistry Matters, 4-7
Counting on Coins, K-2
Cycles of Knowing and Growing, 1-3
Crazy about Cotton, 3-7
Critters, 2-5
Earth Book, 6-9
Electrical Connections, 4-9
Exploring Environments, K-6
Fabulous Fractions, 3-6
Fall into Math and Science, K-1
Field Detectives, 3-6
Finding Your Bearings, 4-9
Floaters and Sinkers, 5-9
From Head to Toe, 5-9
Fun with Foods, 5-9
Glide into Winter with Math and Science, K-1
Gravity Rules! 5-12
Hardhatting in a Geo-World, 3-5
It's About Time, K-2
It Must Be A Bird, Pre-K-2
Jaw Breakers and Heart Thumpers, 3-5
Looking at Geometry, 6-9
Looking at Lines, 6-9
Machine Shop, 5-9
Magnificent Microworld Adventures, 5-9
Marvelous Multiplication and Dazzling Division, 4-5
Math + Science, A Solution, 5-9
Mostly Magnets, 3-6
Movie Math Mania, 6-9
Multiplication the Algebra Way, 6-8
Off the Wall Science, 3-9
Out of This World, 4-8
Paper Square Geometry:
    The Mathematics of Origami, 5-12
Puzzle Play, 4-8
Pieces and Patterns, 5-9
Popping With Power, 3-5
Positive vs. Negative, 6-9
Primarily Bears, K-6
Primarily Earth, K-3
Primarily Magnets, K-2
Primarily Physics, K-3
Primarily Plants, K-3
Primarily Weather, K-3
Problem Solving: Just for the Fun of It! 4-9
Problem Solving: Just for the Fun of It! Book Two, 4-9
Proportional Reasoning, 6-9
Ray's Reflections, 4-8
Sensational Springtime, K-2
Sense-Able Science, K-1
The Sky's the Limit, 5-9
Soap Films and Bubbles, 4-9
Solve It! K-1: Problem-Solving Strategies, K-1
Solve It! 2nd: Problem-Solving Strategies, 2
Solve It! 3rd: Problem-Solving Strategies, 3
Solve It! 4th: Problem-Solving Strategies, 4
Solve It! 5th: Problem-Solving Strategies, 5
Solving Equations: A Conceptual Approach, 6-9
Spatial Visualization, 4-9
Spills and Ripples, 5-12
Spring into Math and Science, K-1
Statistics and Probability, 6-9
Through the Eyes of the Explorers, 5-9
Under Construction, K-2
Water Precious Water, 2-6
Weather Sense: Temperature, Air Pressure, and Wind, 4-5
Weather Sense: Moisture, 4-5
Winter Wonders, K-2

Spanish Supplements*
Fall Into Math and Science, K-1
Glide Into Winter with Math and Science, K-1
Mostly Magnets, 2-8
Pieces and Patterns, 5-9
Primarily Bears, K-6
Primarily Physics, K-3
Sense-Able Science, K-1
Spring Into Math and Science, K-1

* Spanish supplements are only available as downloads from the AIMS website. The supplements contain only the student pages in Spanish; you will need the English version of the book for the teacher's text.

Spanish Edition
Constructores II: Ingeniería Creativa Con Construcciones LEGO® 4-9
    The entire book is written in Spanish. English pages not included.

Other Publications
Historical Connections in Mathematics, Vol. I, 5-9
Historical Connections in Mathematics, Vol. II, 5-9
Historical Connections in Mathematics, Vol. III, 5-9
Mathematicians are People, Too
Mathematicians are People, Too, Vol. II
What's Next, Volume 1, 4-12
What's Next, Volume 2, 4-12
What's Next, Volume 3, 4-12

For further information, contact:
AIMS Education Foundation • P.O. Box 8120 • Fresno, California 93747-8120
www.aimsedu.org • 559.255.6396 (fax) • 888.733.2467 (toll free)

Duplication Rights

Standard Duplication Rights

Purchasers of AIMS activities (individually or in books and magazines) may make up to 200 copies of any portion of the purchased activities, provided these copies will be used for educational purposes and only at one school site.

Workshop or conference presenters may make one copy of a purchased activity for each participant, with a limit of five activities per workshop or conference session.

Standard duplication rights apply to activities received at workshops, free sample activities provided by AIMS, and activities received by conference participants.

All copies must bear the AIMS Education Foundation copyright information.

Unlimited Duplication Rights

To ensure compliance with copyright regulations, AIMS users may upgrade from standard to unlimited duplication rights. Such rights permit unlimited duplication of purchased activities (including revisions) for use at a given school site.

Activities received at workshops are eligible for upgrade from standard to unlimited duplication rights.

Free sample activities and activities received as a conference participant are not eligible for upgrade from standard to unlimited duplication rights.

Upgrade Fees

The fees for upgrading from standard to unlimited duplication rights are:
- $5 per activity per site,
- $25 per book per site, and
- $10 per magazine issue per site.

The cost of upgrading is shown in the following examples:
- activity: 5 activities x 5 sites x $5 = $125
- book: 10 books x 5 sites x $25 = $1250
- magazine issue: 1 issue x 5 sites x $10 = $50

Purchasing Unlimited Duplication Rights

To purchase unlimited duplication rights, please provide us the following:
1. The name of the individual responsible for coordinating the purchase of duplication rights.
2. The title of each book, activity, and magazine issue to be covered.
3. The number of school sites and name of each site for which rights are being purchased.
4. Payment (check, purchase order, credit card)

Requested duplication rights are automatically authorized with payment. The individual responsible for coordinating the purchase of duplication rights will be sent a certificate verifying the purchase.

Internet Use

Permission to make AIMS activities available on the Internet is determined on a case-by-case basis.

P. O. Box 8120, Fresno, CA 93747-8120
aimsed@aimsedu.org • www.aimsedu.org
559.255.6396 (fax) • 888.733.2467 (toll free)